T0335689

Map Construction Algorithms

Mahmuda Ahmed • Sophia Karagiorgou
Dieter Pfoser • Carola Wenk

Map Construction Algorithms

Mahmuda Ahmed
Department of Computer Science
University of Texas at San Antonio
San Antonio, TX, USA

Sophia Karagiorgou
School of Rural and Surveying Engineering
National Technical University of Athens
Zografou, Greece

Dieter Pfoser
Department of Geography
and Geoinformation Science
George Mason University
Fairfax, VA, USA

Carola Wenk
Department of Computer Science
Tulane University
New Orleans, LA, USA

ISBN 978-3-319-25164-6 ISBN 978-3-319-25166-0 (eBook)
DOI 10.1007/978-3-319-25166-0

Library of Congress Control Number: 2015953323

Springer Cham Heidelberg New York Dordrecht London

Printed on acid-free paper

Springer International Publishing AG Switzerland is part of Springer Science+Business Media (www.springer.com)

To

My parents, Jamiul and Raif

– Mahmuda

My loved ones

– Sophia

Nektaria, Daphne, and Alexandros

– Dieter

Joe, Dagmar, and K.-U.

– Carola

Preface

Street maps and transportation networks are of fundamental importance in a wealth of applications. In the past, the production of street maps required expensive field surveying and labor-intensive post-processing. Proprietary data vendors such as NAVTEQ (now Nokia), TeleAtlas (now TomTom), and Google therefore dominated the market. In recent years volunteered geographic information (VGI) efforts such as OpenStreetMap (OSM) have complemented commercial map datasets. They provide map coverage especially in areas that are of less commercial interest. VGI efforts however still require dedicated users to author maps using specialized software tools. Lately on the other hand, the commoditization of GPS technology, its integration in mobile phones, and the advent of low-cost fleet management and positioning applications have triggered the generation of *vast amounts of tracking data*. As a size indicator one can consider the contribution of tracking data in OpenStreetMap, which is steadily increasing and currently amounts to 2.6 trillion points. Besides the use of such data in traffic assessment and forecasting, i.e., map-matching vehicle trajectories to road networks to obtain travel times, there has been a recent surge of actual *map construction algorithms* that derive not only travel time attributes but actual road network geometries from tracking data.

The ambition of this book is to provide the reader with an introduction to map construction algorithms. Providing a research overview is a challenging task since map construction is a very active research field. We address this conundrum by identifying and focusing on three emerging categories of map construction algorithms. For each category we present the general algorithmic idea and a high-level description of the respective algorithms. For this book to also serve as a starting point for map construction research, an in-depth discussion of relevant algorithms is essential. Here, we selected three respective methods, one for each category of algorithms. Devoting one chapter per method, we provide a detailed description that can serve as a basis for subsequent research.

A major challenge in the research community is to compare the performance and to evaluate the quality of competing algorithms. The outcome of map construction is a map dataset that should be close to an actual map data geometry. The quality of an algorithm can thus be measured by the accuracy of its respective result.

Visual inspection has been the most common evaluation approach throughout the literature since it gives an intuitive way of assessing the quality of a map. Parts of this book are dedicated to showcasing map construction results from different algorithms to provide the reader with a simple means to assess the strengths and weaknesses of map construction. Only a few recent works incorporate quantitative distance measures to assess the quality of map construction results. The cross-comparison of different algorithms remains rare, since algorithms and constructed maps are generally not publicly available. In addition, there is a lack of benchmark data and the quantitative evaluation with suitable distance measures is in its infancy. This book discusses the range of existing methods to assess the quality of the constructed maps. These methods are not only discussed in terms of their theoretical characteristics but are also used with different tracking datasets to quantify the quality of the produced maps.

The datasets used in the evaluation were created by tracking vehicle fleets in three large cities. We used datasets from different cities to cover diverse roads (i.e., highways and secondary roads), different sampling rates, and different scales.

In addition to providing a comprehensive comparison of map construction algorithms, we make the mentioned datasets, map construction algorithms and outputs, as well as the evaluation methods publicly available on the Internet at http://www.mapconstruction.org/. We have established this Web site as a repository for map construction data and algorithms, and we invite other researchers to contribute by uploading code and benchmark data supporting their map construction work. We expect that such a central repository will encourage a culture of sharing and will enable the development of improved map construction algorithms.

Organization of This Book

This book seeks to outline the basic principles of map construction algorithms. It deals with the concepts, techniques, and specifically algorithms that have been developed in recent years. An introductory chapter is the basic reference point for all types of readers including practitioners, scientists, and graduate students. The reader will gain an overview of the research ambition so as to also assess the potential of map construction algorithms in her respective field. Beyond covering basic categorization and overview, subsequent chapters give an in-depth analysis of specific techniques. This discussion of map construction algorithms as well as evaluation methods targets researchers interested in advancing the field. Those chapters in connection with the accompanying Web page http://www.mapconstruction.org/ allow for a quick assessment of the state of the art in map construction. The reader is able to download source code and run the algorithms using provided example datasets based on instructions detailed in a user guide chapter in this book. The interested reader will find, at the end of each chapter, a section devoted to bibliographic notes. The book is organized in nine chapters, whose content is as follows.

- Chapter 1 gives an overview of map construction algorithms and groups them into three main categories.
- Chapter 2 describes the TraceBundle algorithm as a proponent of intersection linking algorithms in detail.
- Chapter 3 gives a description of an incremental track insertion algorithm that utilizes the Fréchet distance.
- Chapter 4 presents an example of a density-based map construction algorithm and showcases the use of a pipeline with multiple intermediate steps.
- Chapter 5 visualizes a range of trajectory datasets, reference maps, and map construction results.
- Chapter 6 introduces a range of methods to assess the quality of the constructed maps.
- Chapter 7 is devoted to an experimental evaluation and to establish general performance characteristics of the three algorithms.
- Chapter 8 discusses nontraditional uses for map construction algorithms, i.e., scenarios in which constructing a "map" would provide further insight into the data.
- Chapter 9 provides a user guide for the three map construction algorithms described in detail in this book. The guide shows how to use the actual code and to produce maps based on included trajectory data.

The book has several potential audiences. The first audience includes interested *practitioners* from the geospatial data management community trying to use map construction as a means to simplify and aggregate trajectory datasets. This work will have respective data mining algorithms as its ultimate goal, i.e., to perform data analysis on massive amounts of trajectory data. Another audience consists of *graduate students and researchers* interested in extending the current state of the art in map construction research. This includes, for example, computational geometry researchers pursuing the map construction challenge from a theoretical perspective and aiming for algorithms that provide quality guarantees. We tried as much as possible to cater to both audiences by having overview and example chapters as well as an in-depth discussion of specific methods and specific experimental results. Practitioners will be more interested in Chaps. 1, 5, 8, and 9, which provide an overview of map construction algorithms, visualize some results, discuss new application areas, and provide a user guide, respectively. A more detailed discussion of the algorithms is provided in Chaps. 1–4, with evaluation measures and experimental results being discussed in Chaps. 6 and 7, respectively. In addition, the dedicated Web page and respective user guide of Chap. 9 should allow the reader to start working with the various algorithms right away.

Acknowledgments

Writing this book would not have been possible without the contribution of many people, whom we wish to thank here.

It would not have been possible to develop many of the aspects presented here without common reflection with colleagues, some of them being coauthors of publications. We wish to thank Roy Anati, Maike Buchin, Tom Bylander, Brittany Fasy, Matt Gibson, Mike Goodchild, Kyle Hickmann, Mario Nascimento, Matthias Renz, Kay Robbins, Craig Robinson, Jianhua Ruan, Jessica Sherette, and Andreas Züfle. We especially want to thank James Biagioni for providing the source code to compute the graph sampling-based distance. We also wish to acknowledge the programs and host institutions who supported the research that ultimately lead to this book. This work has been partially supported by the National Science Foundation grant CCF-1301911 and the EU FP7 Marie Curie Initial Training Network GEOCROWD (FP7-PEOPLE-2010-ITN-264994). We are grateful to the members of the GEOCROWD project for the stimulating discussions in the past four years. We are indebted to our home institutions for their continuing support: the Department of Computer Science, University of Texas at San Antonio; the Institute for the Management of Information Systems, "ATHENA" Research Center, Greece; the Department of Geography and Geoinformation Science, George Mason University; and the Department of Computer Science, Tulane University.

We are grateful to our colleagues in these institutions for the warm atmosphere in which this work was carried out, including notably Peggy Agouris, Sotiris Brakatsoulas, Arie Croitoru, Andrew Crooks, Theodoris Dalamagas, Alexandros Efentakis, Tim Leslie, Matt Rice, Timos Sellis, Dimitris Skoutas, Anthony Stefanidis, Iffat Chowdhury, Mohammad Shahedul Islam, Anastasia Kurdia, Mike Mislove, Ram Mettu, Brent Venable, and Qing Wang. We owe a great deal to our administrative staff Nelly Apostolopoulou, Samantha Crooks, Emmy Daniels, Debbie Hutton, and Vanessa Acosta, Susan Allen, Brian Hogan, Cindy Murphy, Debbie Ramil, Pik Rodriguez, and Dan Smolenski for their constant support in everyday life. We also thank our students Christodoulos Efstathiades, Selçuk Karakoç, George Lamprianidis, Sushovan Majhi, and George Skoumas.

Finally, we are grateful for the assistance we received from the editorial staff at Springer. We are especially thankful to Jennifer Malat and Susan Lagerstrom-Fife for their patience during the entire process.

San Antonio, TX, USA	Mahmuda Ahmed
Athens, Greece	Sophia Karagiorgou
Fairfax, VA, USA	Dieter Pfoser
New Orleans, LA, USA	Carola Wenk
July 2015	

Contents

Chapter 1
Map Construction Algorithms

Abstract Map construction methods automatically produce and/or update street map datasets using vehicle tracking data. Enabled by the ubiquitous generation of geo-referenced tracking data, there has been a recent surge in map construction algorithms coming from different computer science domains. This chapter gives a comprehensive overview and comparison of the various algorithms by identifying and focusing on three emerging categories of map construction algorithms. For each category, the general algorithmic idea and a high-level description of the respective algorithms are presented. The overview is complemented by a detailed discussion of several representative algorithms in the following chapters.

1.1 Introduction

The production of street maps requires expensive field surveying and labor-intensive post-processing. Over the last several years, volunteered geographic information (VGI) [18] efforts such as OpenStreetMap (OSM) [20, 27] have complemented commercial map datasets. Still, they require dedicated users to author maps using specialized software tools. At the same time, the ubiquity of GPS positioning combined with the wealth of mobile applications generates now vast amounts of tracking data. As a size indicator one can consider the contribution of tracking data in OpenStreetMap, which currently amounts to 2.6 trillion points [28]. Besides the use of such data in traffic assessment and forecasting [15], i.e., map-matching vehicle trajectories to road networks to obtain travel times [7], there has been a recent surge of actual *map construction algorithms* that derive not only travel time attributes but actual road network geometries from tracking data, e.g., [1–3, 5, 6, 8, 9, 11–14, 16, 17, 19, 21–23, 25, 31–33, 35, 36]. A comprehensive comparison of map construction algorithms that also includes an experimental quality assessment can be found in [4]. An example of a constructed map is given in Fig. 1.1, which shows (a) the vehicle trajectories collected for Berlin in gray color and (b) the respective constructed map using the algorithm of [22] in black color overlayed on an OpenStreetMap background map shown in gray color.

This chapter gives an overview of the state-of-the-art of map construction algorithms. Sketches of some representative methods are provided for three categories of algorithms: (1) Point Clustering, (2) Incremental Track Insertion, and

M. Ahmed et al., *Map Construction Algorithms*, DOI 10.1007/978-3-319-25166-0_1

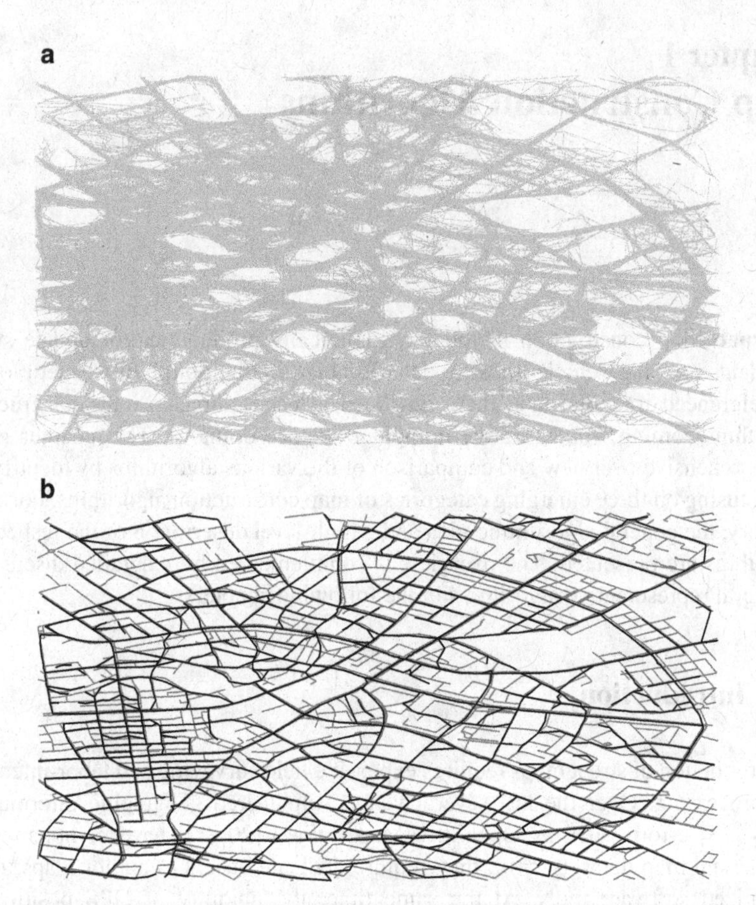

Fig. 1.1 Vehicle tracking data vs constructed map overlaid on ground-truth. (**a**) Vehicle tracking data—*Berlin*. (**b**) Constructed map (in *black*) overlaid on ground-truth (in *gray*)

(3) Intersection Linking. The algorithms discussed include the recent algorithms by Ahmed and Wenk [3], Ge et al. [17], Karagiorgou and Pfoser [22], Cao and Krumm [9], Davies et al. [13], Edelkamp and Schrödl [14], and Biagioni and Eriksson [6]. Subsequent chapters focus one the details of three specific methods, one for each category.

The *map construction problem* is defined as follows: Given a set of input *trajectories*, where each trajectory is a sequence of *measurements*. Each measurement consists of a point (latitude/longitude or (x, y)-coordinates after suitable projection), a time stamp, and optionally additional information such as vehicle heading or speed. The desired output is to construct a *street map*, which is generally modeled as some sort of an embedded graph.

1.2 A Word on Maps

Depending on the desired application and granularity, there are many possible models for street maps. The basic model is a *geometric graph*, where vertices describe intersections and edges represent streets. Typically, the graph is embedded in the plane and it is often assumed to be planar (although this does not model bridges). A common model is an *undirected geometric graph*, where each vertex is embedded as a point in the plane and each edge is a polygonal curve that connects two vertices.

Depending on the application, an intersection can be modeled as a single vertex embedded as a point in the plane, or it could be a set of vertices possibly annotated with turn restrictions, or it could be a region. An edge can be modeled as an abstract connection between vertices, as a (polygonal) curve embedded in the plane, as a set of curves to model multiple lanes, and an edge might be directed to model one-way streets.

The map construction algorithms in the literature generally model the maps as undirected embedded planar graphs or different variants of directed graphs. But often, an undirected graph is computed as a first step and additional information such as edge directions, number of lanes, turn restrictions, and mean speed are computed in an additional post-processing step, e.g. [6, 9, 13, 14, 31].

Later chapters, cf. Chap. 6 on quality measures for map comparison and Chap. 7 for experiments and comparisons, focus on a common street map representation based on undirected embedded graphs, although some algorithms may produce some additional attributes.

1.3 Types of Map Construction Algorithms

There exist several different approaches in the literature for constructing street maps from tracking data. These can be organized into the following categories: (1) *Point clustering*, which includes *k*-means algorithms, density-based methods, and approaches based on neighbourhood complexes; (2) *Incremental track insertion*; and (3) *Intersection linking*. The different approaches, as well as representative algorithms for each category, are described in the following.

1.4 Point Clustering

Map construction algorithms in this general category assume that the input is a set of points, and the points are then clustered in different ways to obtain intersections or street segments that together describe the overall street map. The input point set either comprises the set of all raw input measurements, or a dense sample of all input trajectories. Here, the input trajectories are assumed to be continuous curves obtained from interpolating (usually piecewise-linearly) between measurements.

The point clustering techniques can be reduced to three types of methods for constructing a street map. One approach (cf. [14]) initially clusters the points to generate intersections and then computes the connecting segments as centerlines based on the trajectory points connecting the respective intersection clusters. Other approaches, such as density-based methods, compute the street map in one sweep. The set of points are interpreted as a skeleton image of the road network. The street map is computed as the set of centerlines derived from this image using, e.g., kernel density estimates.

The first approach type, spearheaded by Edelkamp and Schrödl [14], employs the k-means algorithm to cluster the input point set, using distance measures (e.g., Euclidean distance) and possibly also vehicle heading of the measurement, as a condition to introduce seeds at fixed distances along a path. Their map construction algorithm incorporates new algorithms for road segmentation, map-matching, and lane clustering. In [31] this approach was used to refine an existing map rather than building it entirely from scratch. In their short paper [19], Guo et al. make use of statistical analysis of GPS trajectories to extract a center line representation of a street, assuming that the GPS data follows a symmetric 2D Gaussian distribution. This assumption may become unrealistic, especially in error-prone environments. Worrall et al. [35] compute point clusters based on location and heading, and in a second step link these clusters together using non-linear least-squares fitting. They emphasize compression of the input trajectories to infer a digitized road map and present their results only for small datasets. They are mostly concerned with topological elements and not with connected way points. Agamennoni et al. [2] present a machine-learning method to consistently build a representation of the map mostly in dynamic environments such as open-pit mines. They focus on estimating a set of principal curves from the input trajectories to represent the constructed map. Liu et al. [25] first cluster line segments based on proximity and direction, and then use the resulting point clusters and fit polylines to them, to extract road segments.

Another approach first transforms the input point set into a discretized density, using, for example, kernel density estimates. Most of these density-based algorithms function well either when the data is frequently sampled (i.e., once per second) [11], or when there is a lot of redundancy [6, 13, 32, 33]. A similar approach to [6] is presented in Liu et al. [25]. Generally, density-based algorithms have a hard time overcoming the problem of noisy samples when they accumulate in an area. Recently, Wang et al. [34] addressed the problem of map updates by applying their approach to OpenStreetMap data using a density-based approach.

In the computational geometry community, map construction algorithms have been proposed that cluster the input points using local neighborhood properties by employing Voronoi diagrams, Delaunay triangulations [12], or other neighborhood complexes such as the Vietoris-Rips complex [1, 10, 17]. All these algorithms assume a densely sampled input point set, and they provide theoretical quality guarantees for the constructed output map under certain assumptions on the underlying street map and the input trajectories. Aanjaneya et al. [1] view street maps as metric graphs, and they focus on computing the combinatorial structure by computing an almost isometric space with lower complexity. However, they do not compute an explicit embedding of vertices and edges. Chen et al. [12] focus on detecting "good" street portions in the map and connect them subsequently. The theoretical quality guarantees, however, assume dense point sample coverage and error bounds and make assumptions on the road geometry. Ge et al. [17] and Chazal et al. [10] employ the Reeb graph to reconstruct the branching structure. This tool allows the authors to provide topological quality guarantees. Chazal et al. [10] also use metric graphs to model the problem, and they formulate quality guarantees in terms of the Gromov-Hausdorff distance between metric spaces.

1.4.1 Biagioni and Eriksson

Biagioni and Eriksson [6] describe a map construction pipeline that begins with a point clustering-based algorithm that uses density estimation. Their algorithm processes the density function at various thresholds to compute successive versions of an undirected skeleton graph. Subsequently they map-match the input trajectories onto the skeleton in order to clean up noise and to add additional information to the graph, replacing undirected edges with one directed edge per direction, as well as turn lanes at vertices. Just as in [6], for evaluation purposes in Chap. 6 the final undirected topology-refined graph is used as the output. An example output is shown in Fig. 1.2.

1.4.2 Davies et al.

The algorithm by Davies et al. [13] is a classical density-based map construction algorithm. It first computes for each grid cell the density of trajectories that pass

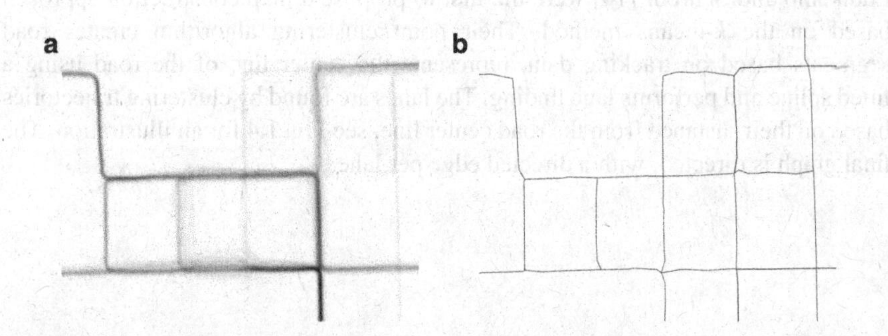

Fig. 1.2 Density-based map construction output after refinement. (**a**) Density. (**b**) Topology-refined graph

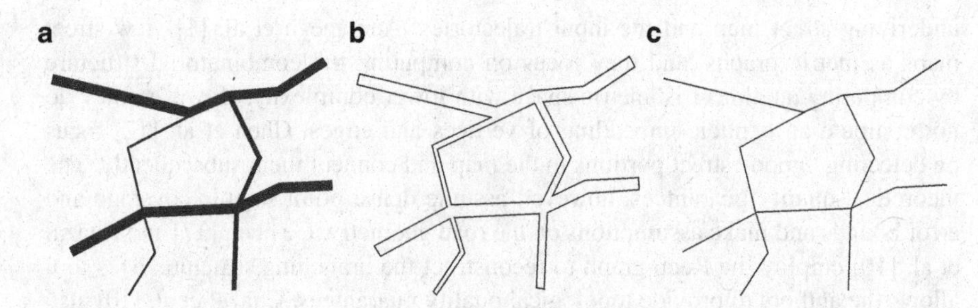

Fig. 1.3 Density-based map construction algorithm. (**a**) Blurred trajectory histogram. (**b**) Contours. (**c**) Centerlines, graph

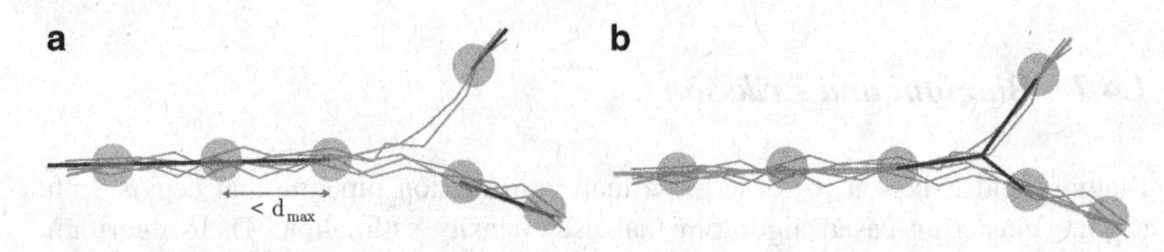

Fig. 1.4 Point clustering-based map construction algorithm—adapted from [14]. (**a**) Input trajectories, clusters, and segments. (**b**) Centerlines, refined graph

through it (cf. the example of Fig. 1.3a). Then it computes the contour of the resulting bit map (Fig. 1.3b), and it uses the Voronoi diagram of the contour to compute a center line representation followed by additional cleanup and assignment of edge directions (Fig. 1.3c). The final output is a directed graph in which each edge is labeled as directed or bidirected.

1.4.3 Edelkamp and Schrödl

Edelkamp and Schrödl [14] were the first to propose a map construction approach based on the *k*-means method. Their point clustering algorithm creates road segments based on tracking data, represents the center line of the road using a fitted spline and performs lane finding. The lanes are found by clustering trajectories based on their distance from the road center line, see Fig. 1.4 for an illustration. The final graph is directed, with a directed edge per lane.

1.4.4 Ge et al.

The algorithm by Ge et al. [17] is a point clustering algorithm that applies topological tools to extract the underlying undirected graph structure. The main idea of this algorithm is to decompose the input data into sets, each corresponding to a single branch in the underlying graph. The authors assume that the input point set is densely sampled, and they compute a two-dimensional neighborhood complex K, such as the Vietoris-Rips complex, on the input points. This requires only the distance matrix of the point set as input. Then, for a given "height" function $f : K \rightarrow \mathbb{R}$, they consider the *Reeb graph* R_f, which intuitively captures a topologically equivalent skeletonization of the complex. Each level set $f^{-1}(r) = \{p \in K \,|\, f(p) = r\}$, for $r \in \mathbb{R}$, may have multiple connected components. The Reeb graph R_f is the quotient space of K where points in the same connected component of a level set are identified. The Reeb graph thus models the connectivity of the connected components of the level sets of f, and hence it captures a one-dimensional graph structure in a natural way.

Figure 1.5 shows an example of a simplicial complex K and a Reeb graph R_f for the height function $f(x, y) = y$. In the paper, the authors consider an intrinsic height function by fixing an arbitrary base point $b \in K$ and defining f to map every point in K to its geodesic distance to b. Finally, there is a canonical way to measure importance of features in the Reeb graph, which allows them to easily simplify the resulting graph. Runtime guarantees are provided, as well as partial quality guarantees for correspondence of cycles. They compared street-maps as sets of cycles. If a cycle in one map does not correspond to a cycle in another map, then obviously a street or a turn is missing in the second map. An embedding for the edges is then obtained by using a principal curve algorithm [24] that fits a curve to the points contributing to the edge.

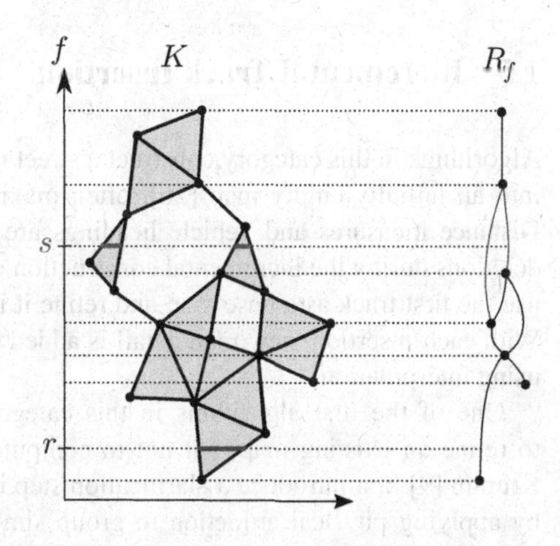

Fig. 1.5 An example of a simplicial complex K and a Reeb R_f with respect to the height function $f(x, y) = y$. The level set $f^{-1}(s)$ has two connected components, while $f^{-1}(r)$ has one connected component, as illustrated by the horizontal segments in the complex K

1.4.5 Aanjaneya et al.

Aanjaneya et al. [1] present a point clustering-based map construction approach that models the problem as a metric reconstruction problem. As input they are given a metric space $(\mathbb{X}_I, d_I)$, which is assumed to be close to an underlying original metric graph $(\mathbb{X}_O, d_O)$. The task is to compute a reconstructed metric graph $(\mathbb{X}_R, d_R)$ that has the same topology as $\mathbb{X}_O$ and a map $\phi : \mathbb{X}_I \to \mathbb{X}_R$ that approximately preserves distances. In practice, the input space $(\mathbb{X}_I, d_I)$ is constructed from the set of input point measurements by computing a weighted local neighborhood graph, such as the one-skeleton of the Vietoris Rips complex. This defines d_I via shortest paths in the graph.

The reconstruction algorithm classifies all input measurements from $\mathbb{X}_I$ into edge points and branch points. This is done by analyzing the number of connected components of the graph $\mathbb{X}_I$ within an annulus around each input measurement; if there are two components it is an edge point, otherwise it is a branch point. Essentially, vertices of the reconstructed metric graph $\mathbb{X}_R$ are formed from connected components of branch points, and edges are formed from connected components of edge points. Here, the metric d_R is defined by assigning to each edge the diameter of the connected component. The authors then prove that if $\mathbb{X}_I$ is close to an original metric graph $\mathbb{X}_O$ then the reconstructed $\mathbb{X}_R$ is homeomorphic to $\mathbb{X}_O$ and distances are approximately preserved. These results depend on several input parameters as well as on several assumptions, including a dense sampling of $\mathbb{X}_I$ and an approximate correspondence between $\mathbb{X}_I$ and $\mathbb{X}_O$. As such, the authors provide the first quality guarantees for a map construction algorithm. The model of the graphs, however, is an undirected graph that has a distance associated with each edge, but interestingly an embedding for the vertices and edges is not explicitly computed, although it can be extracted from the algorithm.

1.5 Incremental Track Insertion

Algorithms in this category construct a street map by incrementally inserting tracks into an initially empty map [26], often making use of map-matching ideas [29]. Distance measures and vehicle headings are also used to perform additions and deletions during the incremental construction of the map. Intuitively, these methods use the first track as a base map and refine it incrementally by adding more tracks. With each insertion, new map detail is added and the existing geometry is updated using interpolation.

One of the first algorithms in this category [30] clusters the tracks in order to refine an existing map, but not to compute a new map from scratch. Cao and Krumm [9] first introduce a clarification step in which they modify the input tracks by applying physical attraction to group similar input tracks together. Then they incrementally insert each track by using local criteria such as distance and direction.

More details of their algorithm are provided in Sect. 1.5.2. Bruntrup et al. [8] propose a spatial clustering-based algorithm that requires high quality tracking data (sampling rate and positional accuracy). The work in [36] discusses a map update algorithm based on spatial similarity. It uses a method similar to GPS trace merging to continuously refine existing road maps. Ahmed and Wenk [3] present an incremental method that employs the Fréchet distance to partially match the tracks to the map. The algorithm is described in more detail in Sect. 1.5.1.

1.5.1 Ahmed and Wenk

The algorithm by Ahmed and Wenk [3] is a simple and practical incremental track insertion algorithm. It models the street map as an undirected embedded graph and uses one parameter ε to model the error associated with the GPS trajectories and with the street width. The insertion of one trajectory proceeds in three steps. The first step performs a partial map-matching of the trajectory to the partially constructed map in order to identify matched portions and unmatched portions, see Fig. 1.6a for an example. This partial map-matching is based on a variant of the Fréchet distance. For matched portions the Fréchet distance of the sub-curve to the map is at most ε, and for unmatched portions the Fréchet distance of the sub-curve to the map is greater than ε.

In the second step, the unmatched portions of the trajectory are then inserted into the partially constructed map, which requires creating new vertices and creating and splitting edges. In a third step, the already existing edges in the map, that are covered by the matched portions of the trajectory, are updated using a minimum-link algorithm to compute a new representative edge. See Fig. 1.6b for an example of the graph after inserting the trajectory. Note that the last step is only needed to provide a guaranteed bound on the complexity of the output map; in the implementation of this algorithm that is used in Chap. 7, this last step has been omitted. Ahmed and Wenk also give theoretical quality guarantees for the output map computed by

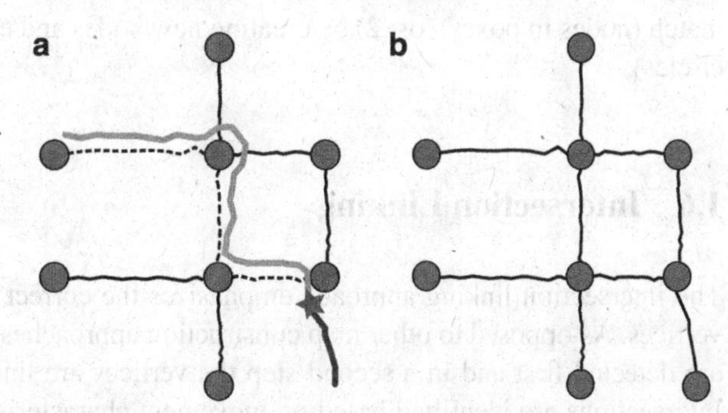

Fig. 1.6 Incremental track insertion algorithm. The matched portions of the trajectory are shown in *lighter shade*, and the unmatched portions in *darker shade*. (**a**) Existing graph and trajectory to be inserted. (**b**) Graph after inserting the trajectory

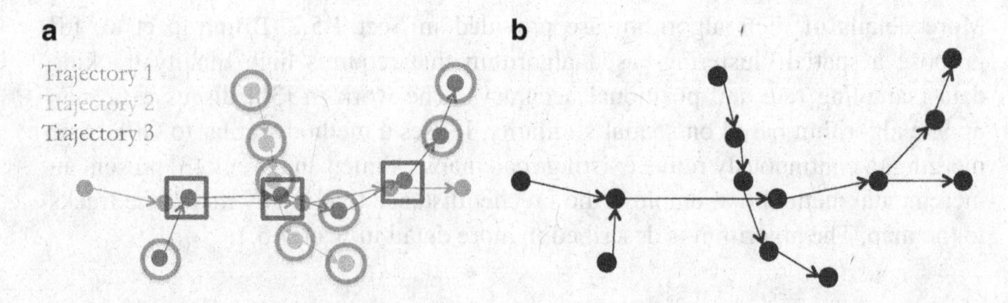

Fig. 1.7 The incremental track insertion algorithm—adapted from [9]. (**a**) Three input trajectories. (**b**) Merged graph

their algorithm, which include a one-to-one correspondence between well-separated "good" portions of the underlying map and the output map, with a guaranteed Fréchet distance between those portions.

1.5.2 Cao and Krumm

This incremental track insertion approach models the street map as a directed embedded graph with one directed edge per direction. It proceeds in two stages. In the first stage, simulation of physical attraction is used to modify the input tracks to group portions of the tracks that are similar together. This results in a cleaner dataset in which track clusters are more pronounced and the two differently directed portions of a road segment are more separated. This much cleaner data is then used as the input for a fairly simple incremental track insertion algorithm. This algorithm makes local decisions based on distance and direction to insert an edge or vertex and either merges the vertex onto an existing edge, or adds a new edge and vertex.

Figure 1.7 gives a respective map construction example. The three trajectories of Fig. 1.7a are used to incrementally build the graph in Fig. 1.7b by (1) either merging nodes to existing nodes if the distances are small and the directions of the trajectories match (nodes in boxes), or (2) by creating new nodes and edges otherwise (nodes in circles).

1.6 Intersection Linking

The intersection linking approach emphasizes the correct detection of intersection vertices. As opposed to other map construction approaches, the intersection vertices are detected first and in a second step the vertices are linked together with edges. Intersections are identified based on movement characteristics (speed, direction) or point density. The intersections are then linked by interpolating the geometry of the connecting trajectories. Fathi and Krumm [16] provide an approach that detects intersections by using a prototypical detector trained on ground-truth data from an existing map. This approach works best for frequently sampled data (1–5 s) and grid-like road networks.

The method by Karagiorgou and Pfoser [22] relies on detecting changes in the direction of movement to infer intersection nodes, and then "bundling" the trajectories around them to create the map edges. It uses less frequently sampled data (> 30 s) and produces street maps for arbitrary road network geometries.

1.6.1 Fathi and Krumm

The map construction algorithm by Fathi and Krumm [16] was one of the first to construct a map from intersection nodes. An intersection is a location where more than two edges connect to each other. To detect intersections, one has to determine if an arbitrary location on the map is an intersection or not by examining closeby GPS trajectories. This approach uses a shape descriptor that can discriminate between intersections and non-intersections and that can be represented as a feature vector. The shape descriptor captures the local distribution and direction of trajectory edges in a circle around all candidate trajectory points. The circle consists of a set of annular sections to essentially rasterize the incident edges with respect to the trajectory point. Each section of the circle can be thought of as a histogram bin. For each incident edge, a point is added to the bin the edge passes through. Mapping the bins of each shape descriptor to a vector, one can compute a classifier based on all the feature vectors provided by the training examples, i.e., from a known road network. Roads connecting intersections are derived from trajectories by choosing the minimum-distance trajectory connecting the two intersection nodes. The locations of the detected intersections are further refined by iteratively matching the model (detected intersections and roads) to the data (GPS trajectories) to obtain a transformation that allows one to readjust the model and to obtain preciser road map geometries. The authors show that the detected intersections deviated from their ground-truth intersections by approximately 4.6 m.

1.6.2 Karagiorgou and Pfoser

The algorithm by Karagiorgou and Pfoser [22] is a heuristic approach that "bundles" trajectories around intersection nodes. It represents the street map as a directed embedded graph in which each edge is labeled as directed or bi-directed.

The main contribution of the so-called *TraceBundle* algorithm is its methodology to derive intersection nodes. It relies on detecting changes in movement to cluster "similar" nodes. A node at which a change in direction and speed occurs is considered a turn indicator. Turn clusters are then produced based on (1) the

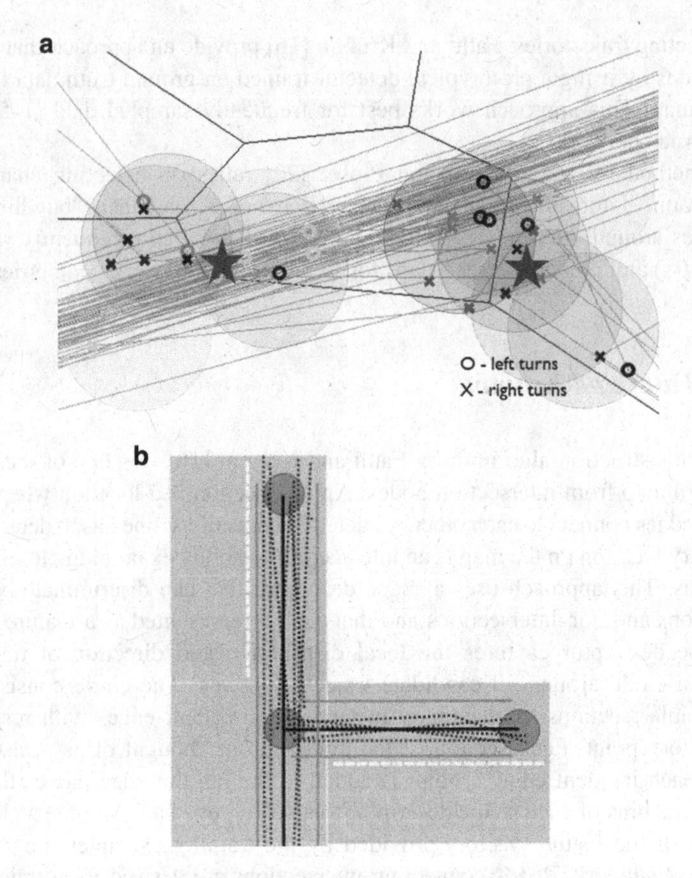

Fig. 1.8 The *TraceBundle* algorithm [22]. (**a**) Intersection nodes. (**b**) Compacting links

spatial proximity and (2) the turn type of a set of turn indicators. The centroid of a turn cluster then becomes an intersection node. Links between intersection nodes are derived by compacting the trajectories connecting the intersection nodes. Figure 1.8 visualizes the steps of the algorithm. Figure 1.8a shows the constructed intersection nodes as black stars. The constituting turn clusters are shown as *x* and *o* markers. Figure 1.8b shows the links between intersection nodes as black lines. The constituting trajectories are shown as dashed lines.

The TraceBundle algorithm has three tunable parameters: *angular difference*, *speed* and *spatial proximity*. Angular difference is the relative change of the vehicle direction measured in *degrees*. The speed threshold indicating turning vehicles is measured in km/h. This is an empirical maximum threshold to separate high-speed turns from turns at intersections. The spatial proximity distance threshold for clustering turn clusters into intersection nodes is measured in meters.

References

1. Aanjaneya, M., Chazal, F., Chen, D., Glisse, M., Guibas, L.J., Morozov, D.: Metric graph reconstruction from noisy data. In: Proceedings of 27th ACM Symposium on Computational Geometry, pp. 37–46 (2011)
2. Agamennoni, G., Nieto, J.I., Nebot, E.M.: Robust inference of principal road paths for intelligent transportation systems. IEEE Trans. Intell. Transp. Syst. **12**(1), 298–308 (2011)
3. Ahmed, M., Wenk, C.: Constructing street networks from GPS trajectories. In: Proceedings of 20th Annual European Symposium on Algorithms, pp. 60–71 (2012)
4. Ahmed, M., Karagiorgou, S., Pfoser, D., Wenk, C.: A comparison and evaluation of map construction algorithms using vehicle tracking data. GeoInformatica **19**(3), 601–632 (2015)
5. Biagioni, J., Eriksson, J.: Inferring road maps from global positioning system traces: survey and comparative evaluation. Transp. Res. Rec. J. Transp. Res. Board **2291**, 61–71 (2012)
6. Biagioni, J., Eriksson, J.: Map inference in the face of noise and disparity. In: Proceedings of 20th ACM SIGSPATIAL International Conference on Advances in Geographic Information Systems, pp. 79–88 (2012)
7. Brakatsoulas, S., Pfoser, D., Salas, R., Wenk, C.: On map-matching vehicle tracking data. In: Proceedings of 31st VLDB Conference, pp. 853–864 (2005)
8. Bruntrup, R., Edelkamp, S., Jabbar, S., Scholz, B.: Incremental map generation with GPS traces. In: Proceedings of IEEE Intelligent Transportation Systems, pp. 574–579 (2005)
9. Cao, L., Krumm, J.: From GPS traces to a routable road map. In: Proceedings of 17th ACM SIGSPATIAL International Conference on Advances in Geographic Information Systems, pp. 3–12 (2009)
10. Chazal, F., Huang, R., Sun, J.: Gromov-Hausdorff approximation of filament structure using Reeb-type graph. Discret. Comput. Geom. **53**(3), 621–649 (2015)
11. Chen, C., Cheng, Y.: Roads digital map generation with multi-track GPS data. In: Proceedings of Workshops on Education Technology and Training, and on Geoscience and Remote Sensing, pp. 508–511. IEEE (2008)
12. Chen, D., Guibas, L.J., Hershberger, J.E., Sun, J.: Road network reconstruction for organizing paths. In: Proceedings of 21st ACM-SIAM Symposium on Discrete Algorithms, pp. 1309–1320 (2010)
13. Davies, J.J., Beresford, A.R., Hopper, A.: Scalable, distributed, real-time map generation. IEEE Pervasive Comput. **5**(4), 47–54 (2006)
14. Edelkamp, S., Schrödl, S.: Route planning and map inference with global positioning traces. In: Computer Science in Perspective, pp. 128–151. Springer, Berlin (2003)
15. Efentakis, A., Brakatsoulas, S., Grivas, N., Lamprianidis, G., Patroumpas, K., Pfoser, D.: Towards a flexible and scalable fleet management service. In: Proceedings of 6th ACM SIGSPATIAL International Workshop on Computational Transportation Science, pp. 79–84 (2013)
16. Fathi, A., Krumm, J.: Detecting road intersections from GPS traces. In: Proceedings of 6th International Conference on Geographic Information Science, pp. 56–69 (2010)
17. Ge, X., Safa, I., Belkin, M., Wang, Y.: Data skeletonization via Reeb graphs. In: Proceedings of 25th Annual Conference on Neural Information Processing Systems, pp. 837–845 (2011)
18. Goodchild, M.F.: Citizens as voluntary sensors: spatial data infrastructure in the world of web 2.0. Int. J. Spat. Data Infrastruct. Res. **2**, 24–32 (2007)
19. Guo, T., Iwamura, K., Koga, M.: Towards high accuracy road maps generation from massive GPS traces data. In: Proceedings of IEEE International Geoscience and Remote Sensing Symposium, pp. 667–670 (2007)
20. Haklay, M., Weber, P.: OpenStreetMap: user-generated street maps. IEEE Pervasive Comput. **7**(4), 12–18 (2008)
21. Jang, S., Kim, T., Lee, E.: Map generation system with lightweight GPS trace data. In: Proceedings of 12th International Conference on Advanced Communication Technology, pp. 1489–1493 (2010)

22. Karagiorgou, S., Pfoser, D.: On vehicle tracking data-based road network generation. In: Proceedings of 20th ACM SIGSPATIAL International Conference on Advances in Geographic Information Systems, pp. 89–98 (2012)
23. Karagiorgou, S., Pfoser, D., Skoutas, D.: Segmentation-based road network construction. In: Proceedings of 21st ACM SIGSPATIAL International Conference on Advances in Geographic Information Systems, pp. 450–453 (2013)
24. Kégl, B., Krzyzak, A., Linder, T., Zeger, K.: Learning and design of principal curves. IEEE Trans. Pattern Anal. Mach. Intell. **22**(3), 281–297 (2000)
25. Liu, X., Biagioni, J., Eriksson, J., Wang, Y., Forman, G., Zhu, Y.: Mining large-scale, sparse GPS traces for map inference: comparison of approaches. In: Proceedings of 18th ACM SIGKDD International Conference on Knowledge Discovery and Data Mining, pp. 669–677 (2012)
26. Niehofer, B., Burda, R., Wietfeld, C., Bauer, F., Lueert, O.: GPS community map generation for enhanced routing methods based on trace-collection by mobile phones. In: Proceedings of 1st International Conference on Advances in Satellite and Space Communications, pp. 156–161 (2009)
27. OpenStreetMap (2015). http://www.openstreetmap.org/
28. OpenStreetMap Foundation: Bulk GPX track data (2013). http://www.blog.osmfoundation.org/2013/04/12/bulk-gpx-track-data/
29. Quddus, M., Ochieng, W., Noland, R.: Current map-matching algorithms for transport applications: state-of-the art and future research directions. Transp. Res. C Emerg. Technol. **15**, 312–328 (2007)
30. Rogers, S., Langley, P., Wilson, C.: Mining GPS data to augment road models. In: Proceedings of 5th ACM SIGKDD International Conference on Knowledge Discovery and Data Mining, pp. 104–113 (1999)
31. Schroedl, S., Wagstaff, K., Rogers, S., Langley, P., Wilson, C.: Mining GPS traces for map refinement. Data Min. Knowl. Disc. **9**, 59–87 (2004)
32. Shi, W., Shen, S., Liu, Y.: Automatic generation of road network map from massive GPS vehicle trajectories. In: Proceedings of 12th International IEEE Conference on Intelligent Transportation Systems, pp. 48–53 (2009)
33. Steiner, A., Leonhardt, A.: Map generation algorithm using low frequency vehicle position data. In: Proceedings of 90th Annual Meeting of the Transportation Research Board, pp. 1–17 (2011)
34. Wang, Y., Liu, X., Wei, H., Forman, G., Chen, C., Zhu, Y.: Crowdatlas: Self updating maps for cloud and personal use. In: Proceedings of 11th International Conference Mobile Systems, Applications and Services (2013)
35. Worrall, S., Nebot, E.: Automated process for generating digitised maps through GPS data compression. In: Proceedings of Australasian Conference on Robotics and Automation (2007)
36. Zhang, L., Thiemann, F., Sester, M.: Integration of GPS traces with road map. In: Proceedings of 3rd ACM SIGSPATIAL International Workshop on Computational Transportation Science, pp. 17–22 (2010)

Chapter 2
TraceBundle Map Construction Algorithm

Abstract This chapter presents the *TraceBundle* algorithm, which is a representative of the intersection linking category of map construction algorithms. The main approach is to first detect intersection nodes, then "bundle" trajectories around them in order to construct edges. Changes in movement direction and speed are used as turn indicators, and similar turns are combined to form intersection nodes. In an improved version of the algorithm the hierarchical nature of the road network is considered and different road categories are taken into account. By segmenting the trajectories based on speed, hierarchical road network layers are derived which are then combined into a single network. Segmentation also addresses the challenges imposed by noisy, low-sampling rate trajectories and provides for a mechanism for accommodating incremental map updates.

2.1 Introduction

This section presents a map construction algorithm that takes vehicle tracking data in the form of trajectories as input and produces a directed embedded road network graph. The presented *TraceBundle* algorithm emphasizes the correct detection of intersection nodes and the linking of these nodes both in terms of connectivity and actual geometry. Intersections are identified based on movement characteristics (speed, direction) and point density. The intersections are then linked by interpolating the geometry of the connecting trajectories. The name TraceBundle derives from "trace", an alternative label for trajectories, and "bundle" to capture the basic intuition behind the algorithm. TraceBundle bundles redundant trajectory data and can handle road networks of arbitrary geometries. It also extracts road category information for each road segment.

TraceBundle falls into the category of *trace clustering* map construction algorithms. In the literature, algorithms in this category either follow map matching [8] or heuristic approaches by aggregating GPS trajectories into an incrementally-built road network [7]. Distance measures and vehicle heading are also used to perform additions and deletions to the map. An early approach in that respect uses trace clustering to refine an existing road network, but not to actually construct one from scratch [9]. In [2], a method is presented to eliminate noise in GPS traces, while [3] provide an approach that detects intersections by using a prototypical detector

M. Ahmed et al., *Map Construction Algorithms*, DOI 10.1007/978-3-319-25166-0_2

trained on ground-truth data from an existing map. This approach works best for well-aligned road networks and with frequently sampled data, e.g, with a sampling frequency of 5 s. Bruntrup et al. [1] and Liu et al. [6] construct a road network, but require accurate data and high sampling rates (every 1 s). Zhang et al. [10] use a method similar to GPS trace merging to continuously refine existing road maps. The methods proposed in [4, 5] and presented in this chapter differ from the aforementioned approaches in that they focus on preserving the connectivity of the underlying road network. This is achieved by clustering the trajectories based on intersection indicators (turn samples) and speed profiles.

2.2 TraceBundle Algorithm

The following algorithm relies on a heuristic method to detect turns and derive intersections from the trajectory data. The map is constructed by bundling trajectories that connect the same intersection nodes. In addition, road categories are derived based on the amount of data that is available for particular road network portions, i.e., on the traversal frequency. Figure 2.1 plots trajectories with the principal roads of the actual road network being (at least visually) evident.

The algorithm proceeds in three essential steps: (1) *identify intersections*, i.e., use turns in vehicle trajectories as indicators for intersections, (2) *connect intersections*, i.e., create edges between intersections by using trajectories, and (3) *extract the compacted network graph*, i.e., compact the edges to create a meaningful road network graph.

Fig. 2.1 Trajectories of a Berlin taxi fleet

Algorithm 2.1. Identify intersections

Input: A set of trajectories T
Output: A set of intersection nodes I

```
1  begin
2      P ← ∅ ▷ Position samples in one trajectory
3      PS ← ∅ ▷ Turn samples
4      PC ← ∅ ▷ Turn clusters
5      I ← ∅ ▷ Intersection nodes
6      α, v, d_c, d_i ▷ Angle, speed and distance thresholds
7      ▷ Process all position samples in all trajectories
8      while T[i] ≠ null do
9          P ← T[i] ▷ Position samples of a single trajectory
10         a_p ← AngularDiff(P[i − 1], P[i], P[i + 1]) ▷ Angular Difference
11         v_p ← δx(P[i − 1], P[i]) / δt(P[i − 1], P[i]) ▷ Mean speed
12         if a_p > α and v_p < v then
13             PS.insert(P[i], TurnType(P[i])) ▷ Turn sample
14         end
15     end
16     ▷ Cluster turn samples into turn clusters
17     PC ← ClusterTurns(PS, d_c)
18     ▷ Cluster turn clusters into intersection nodes
19     I ← ClusterIntersections(PC, d_i)
20 end
```

2.2.1 *Turns and Intersections*

Given a vehicle trajectory, turns are used to detect intersection nodes of the road network. The position samples are grouped into *turn clusters* by using specific turn indicators that indicate changes of the vehicle's movement in terms of speed and direction. Then the turn clusters are grouped to form intersection nodes. The algorithm to *identify intersections* in the trajectory data is given in Algorithm 2.1.

2.2.1.1 Turn Indicators

From a common-sense understanding of vehicular movement it is evident that when a vehicle turns it (1) reduces its speed and (2) changes its direction. The presented algorithm uses 40 km/h as a *reduced speed indicator* in combination with a change of direction. Figure 2.2 gives an example of a trajectory showing position samples and the respective *direction vectors*. A direction threshold of 15° was experimentally established as a suitable choice for road networks. In Algorithm 2.1, all trajectories are scanned in a position-by-position and an edge-by-edge manner (lines 9–16). All position samples that satisfy both turn conditions (lines 11 and 12) are recorded as *turn samples* (line 14).

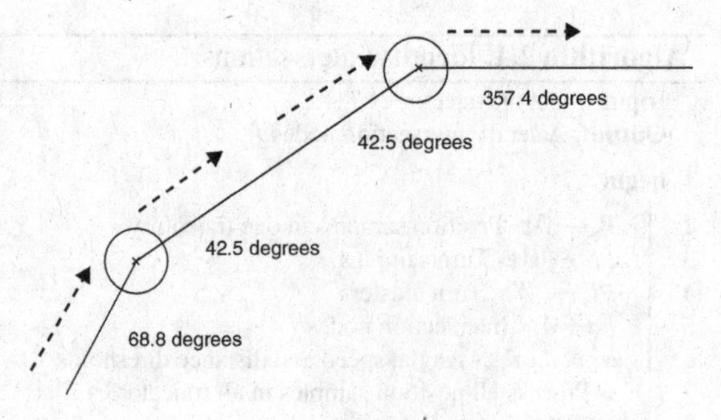

Fig. 2.2 Angular difference

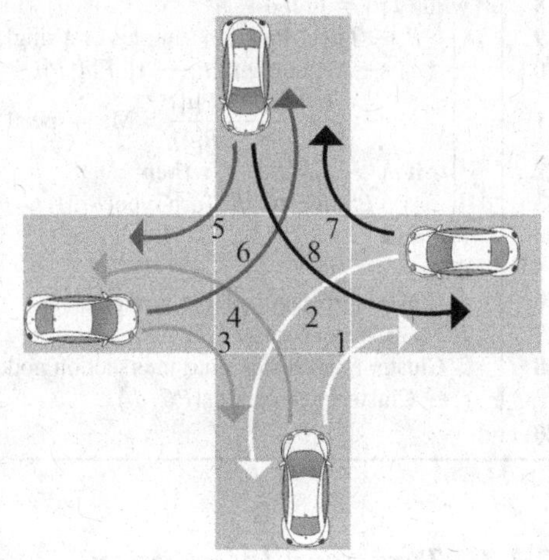

Fig. 2.3 Turn model

2.2.1.2 Clustering Turns

A *turn model* is used to cluster samples from different trajectories in order to derive intersections. The turn model that is used for this algorithm describes all possible movement patterns at an intersection node by the angle with respect to the positive *x*-axis centered at the intersection. The turn samples are classified by using *eight types of turns* as shown in Fig. 2.3.

All discovered turn samples are categorized according to the corresponding turn identifier of the model. The turn samples can then be grouped according to (1) spatial proximity and (2) turn similarity. Choosing a proximity threshold of 50 m, agglomerative hierarchical clustering is used to derive *turn clusters* (cf. Algorithm 2.1, Line 18). The location of a turn cluster is the centroid of the position samples in the cluster. Figure 2.4a shows the computed result for three roads that meet at an intersection.

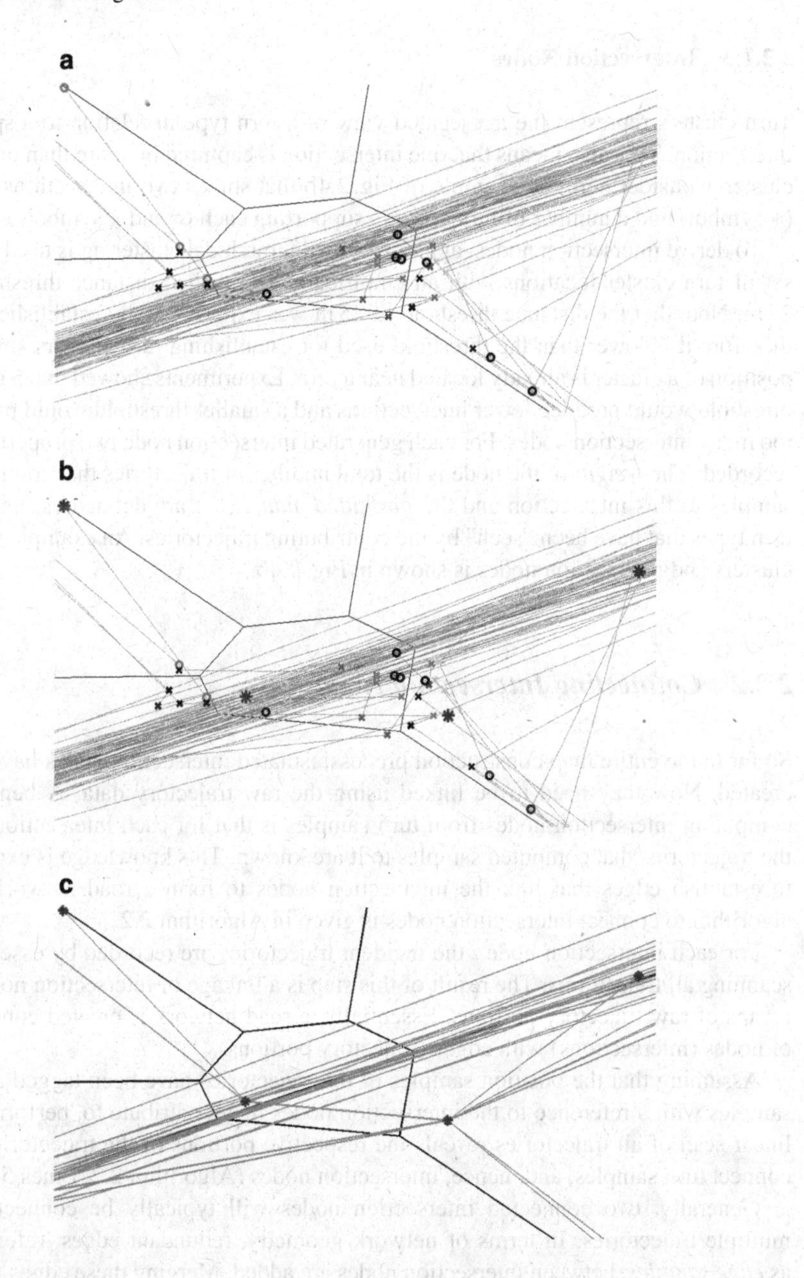

Fig. 2.4 Computing intersection nodes. An odd turn type is indicated by an × symbol and an even turn type by an ∘ symbol. Color is used to distinguish turn types further, *yellow* for 1 and 2, *orange* for 3 and 4, *red* for 5 and 6, and *black* for 7 and 8. Intersection nodes are indicated by a gray $*$. (**a**) Computing turn clusters from position samples. (**b**) Computing intersection nodes from turn clusters. (**c**) Connecting intersection nodes (Color figure online)

2.2.1.3 Intersection Nodes

Turn clusters represent the aggregated view of a turn type in relation to a specific intersection. This also means that one intersection is captured by more than one turn cluster. Consider here the example of Fig. 2.4b that shows two intersections nodes ($*$ symbol) and a number of turn clusters supporting each ($\circ$ and $\times$ symbols).

To derive intersection nodes, agglomerative hierarchical clustering is used on the set of turn cluster locations with an empirically established distance threshold of 25 m. Note that the distance threshold of 25 m was experimentally established, and therefore it is lower than the threshold used for establishing turn clusters since the position of a cluster is already located near a turn. Experiments showed that a greater threshold would produce fewer intersections and a smaller threshold would produce too many intersection nodes. For each generated intersection node two properties are recorded. The *weight* of the node is the total number of trajectories that contributed samples to this intersection and the *permitted maneuvers* are defined as the set of turn types that have been "seen" by the contributing trajectories. An example of turn clusters and intersection nodes is shown in Fig. 2.4b.

2.2.2 *Connecting Intersection Nodes*

So far in the entire map construction process, isolated intersection nodes have been created. Now they need to be linked using the raw trajectory data. A benefit of computing intersection nodes from turn samples is that for each intersection node the trajectories that computed samples to it are known. This knowledge is exploited to establish edges that link the intersection nodes to form a road network. The algorithm to connect intersection nodes is given in Algorithm 2.2.

For each intersection node i the incident trajectories are recorded by essentially scanning all trajectories. The result of this step is a linkage of intersection nodes by means of raw trajectory portions. Essentially, a road network is created consisting of nodes (intersections) with edges (trajectory portions).

Assuming that the position samples of the trajectories have been tagged as turn samples with a reference to the intersection nodes they contribute to, performing a linear scan of all trajectories reveals the respective portions of the trajectories that connect turn samples, and, hence, intersection nodes (Algorithm 2.2, Lines 5–12).

Generally, two connected intersection nodes will typically be connected by multiple trajectories. In terms of network geometry, redundant edges, referred to as *edge samples*, between intersection nodes are added. Merging these edges will be the next step. To compute the geometry of an edge based on the recorded samples, a sweep-line algorithm is used to merge the edge samples. Given a set of edge samples, each consisting of a set of points, for each point an average position based on the normal distance of the points to all other edge samples is computed and the point of the edge sample is translated to this average position. The edge geometry is then the ordered sequence of the adjusted points of all edge samples. Figure 2.5a

Algorithm 2.2. Connect intersection nodes using trajectories

Input: Set of trajectories T and intersection nodes I
Output: A set of edges E

```
1  begin
2  |  ES ← ∅ ▷ Edge samples
3  |  ▷ Identify intersection sequences from trajectories
4  |  foreach t ∈ T do
5  |  |  foreach p ∈ t do
6  |  |  |  foreach i ∈ I do
7  |  |  |  |  ▷ In constituting position sample
8  |  |  |  |  if p ∈ i.samples() then
9  |  |  |  |  |  ▷ Record current node i and previous intersection node i'
10 |  |  |  |  |  ▷ along with connecting trajectory portion
11 |  |  |  |  |  ES.add({i, i', p, ..., p'})
12 |  |  |  |  end
13 |  |  |  end
14 |  |  end
15 |  end
16 |  ▷ Edges = merged edge samples
17 |  E ← ES.sweepMerge()
18 end
```

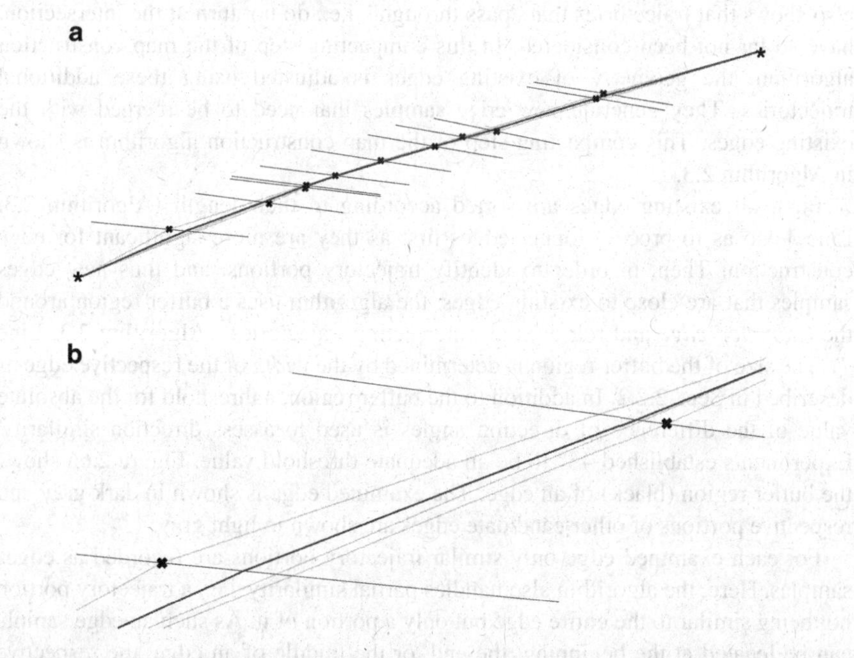

Fig. 2.5 Edge samples. **(a)** Edge samples between two intersections. **(b)** Average positions of edges samples

shows a set of edge samples (gray) that connect two intersection nodes marked by an $*$. The points of the edge samples are marked by an $\times$. Horizontal thin lines indicate points at which the normal distance to edge samples is measured and the average position is computed. Figure 2.5b shows a close up of two edge samples and respective points.

In addition to the geometry, for each edge (1) its *weight* is computed as the number of constituting edge samples and (2) its *width* is derived as the maximum spatial extent of the edge samples. In general, the number of generated intersection nodes depends highly on the parameter setting of the algorithm, i.e., choosing lower or higher threshold values will generate more or fewer intersection nodes, respectively. The parameters need to be tuned to the network and the data in question.

2.2.3 Compacting Edges

So far the constructed road network consists of intersection nodes that are connected by edges, which have been derived from trajectories that turn at these intersections. Figure 2.4c shows how intersection nodes are connected by various trajectories. It also shows that trajectories that "pass through", i.e., do not turn at the intersection, have so far not been considered. In this compacting step of the map construction algorithm, the geometry of existing edges is adjusted using these additional trajectories. They generate new edge samples that need to be merged with the existing edges. This compacting step of the map construction algorithm is shown in Algorithm 2.3.

First, all existing edges are sorted according to their length (Algorithm 2.3, Line 1) so as to process longer edges first as they are more significant for edge construction. Then, in order to identify trajectory portions, and thus new edges samples that are close to existing edges, the algorithm uses a buffer region around the *examined edge* and retrieves all intersecting trajectories. (Algorithm 2.3, Line 7). The size of the buffer region is determined by the *width* of the respective edge as described in Sect. 2.2.2. In addition to the buffer region, a threshold for the absolute value of the difference of direction angles is used to assess direction similarity. Experiments established 45° to be an adequate threshold value. Figure 2.6a shows the buffer region (black) of an edge. The examined edge is shown in dark gray and respective portions of other candidate edges are shown in light gray.

For each examined edge only similar trajectory portions are recorded as edges samples. Here, the algorithm also handles partial similarity, i.e., a trajectory portion not being similar to the entire edge but only a portion of it. As such an edge sample can be located at the beginning, the end, or the middle of an edge, the respective trajectory portion is split (Algorithm 2.3, Line 14).

Algorithm 2.3. Compact edges

Input: Set of existing edges E, set of trajectories T
Output: Updated edges E^*

1 **begin**
2 $E \leftarrow E$.sort() ▷ Sort edges by length
3 $ES \leftarrow \emptyset$ ▷ Edge samples
4 *Angle* ▷ Direction threshold
5 **foreach** $e \in E$ **do**
6 **foreach** $t \in T$ **do**
7 ES.add(e.buffer(t, *Angle*))
8 **foreach** $es \in ES$ **do**
9 **if** e.contains(es) **then**
10 e.update(e.sweepMerge(es))
11 **end**
12 **else**
13 ▷ Partial overlap $es_{in}, es_{out} \leftarrow es.split(e)$
14 ES.add(es_{out})
15 e.update(e.sweepMerge(es_{in}))
16 **end**
17 **end**
18 **end**
19 **end**
20 **end**

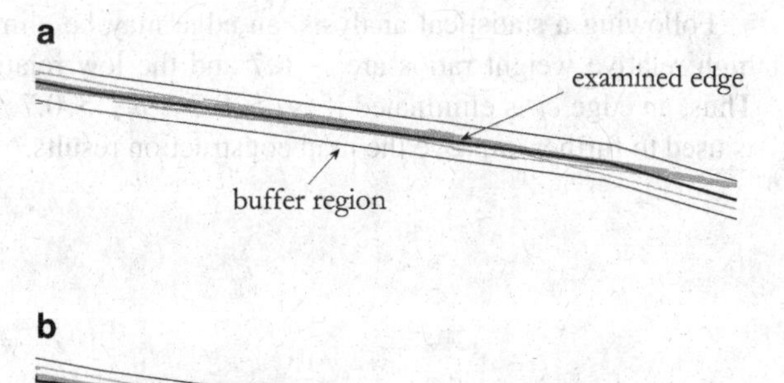

Fig. 2.6 Road network refinement—compacting edges. (**a**) Edge, edge samples and buffer region. (**b**) Updated edge

Finally, edge samples are merged with existing edges following the approach of Sect. 2.2.2. The method is applied to all generated edges samples (partial or whole, Algorithm 2.3, Lines 10 and 16). Updated edges also update their weight and width adding the number and extent of the merged edges samples.

2.2.4 Post-Processing

While the road extraction algorithm so far has already created a road network graph, the following heuristic post-processing step further improves the quality of the result. The basic idea in the presented map construction algorithm is the use of turns to identify turn clusters, which in turn create intersection nodes. Trajectories are recorded by sampling the motion. In the case of turns this is especially critical, in that a position sample might create turn clusters well in advance, or after the actual turn and hence introduce additional intersections. This phenomenon is referred to as *triangular intersections*. To detect such triangular intersections, edge weights are analyzed in relation to geometric properties of the road network. Let e_i and e_j be two edges incident to the same intersection node, and let w_i and w_j be their respective weights. Then $\rho_{i,j} = w_i/w_j$ is defined as their relative weight. Now, the aim is to detect triangular configurations of three edges e_1, e_2, e_3 where two sides have relatively high relative weights in relation to the third side, i.e., $\rho_{1,2} \gg \rho_{1,3}$ and $\rho_{1,2} \gg \rho_{2,3}$, see Fig. 2.7 for an example.

Following a statistical analysis, an edge may be eliminated provided that both high relative weight ratios are > 0.7 and the low relative weight ratio is < 0.6. Thus, an edge e_k is eliminated if $\rho_{i,j} > 0.7 \wedge \rho_{i,k} > 0.7 \wedge \rho_{j,k} < 0.6$. This heuristic is used to further improve the map construction results.

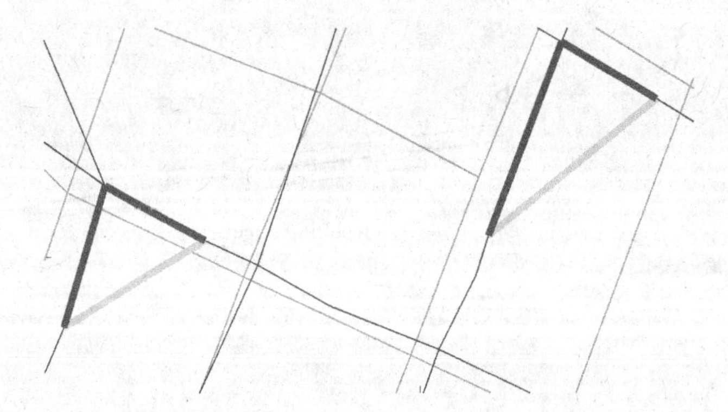

Fig. 2.7 Triangular intersections. Edges with high relative weight are shown in *darker* color (*red*) and with low relative weight are shown in *lighter* color (*yellow*) (Color figure online)

2.3 TraceConflation Algorithm

The basic *TraceBundle* algorithm can be improved when considering the semantics of movement. Not all roads are equal and road categories, besides speed, also determine the actual road geometry. Major roads, also due to higher speeds, facilitate a different type of movement than neighborhood roads. The former has fewer changes in direction than the latter. Considering these aspects of movement, the proposed *TraceConflation* algorithm constructs a road network in a layered fashion based on segmenting the trajectory data using speed categories. The process involves three steps: (1) the *segmentation of trajectories*, i.e., splitting the input dataset of trajectories into subsets of (sub-)trajectories according to their characteristics, (2) the *construction of the network layers*, i.e., processing each subset to identify nodes and edges of the network, and (3) the *conflation of the network layers*, i.e., merging the generated layers to produce the complete map.

2.3.1 Segmentation of Trajectories

A main challenge when inferring a movement network from raw GPS trajectories is that the data is often noisy and heterogeneous (GPS errors, missing values, different sampling rates, different speeds, etc.). Thus, treating all the input data equally, inevitably introduces inaccuracies in the result. This problem is addressed by splitting the trajectories into subsets according to movement speed. In this way each subset can be treated separately, e.g., by refining the parameters of the map inference algorithm accordingly. The aim is to derive different (but probably overlapping) portions of the network with higher accuracy, which then need to be merged in order to produce the complete network. Hence, this process leads to a layered construction of the network.

Three speed categories ("slow", "medium", "fast") are considered and used to classify the trajectories accordingly. Notice that typically an object may have moved with different speeds across different parts of the trajectory. In this case the trajectory needs to be split into sub-trajectories, with each one assigned to the corresponding speed category subset. To categorize the trajectories, first a *speed* value is assigned to each segment of a trajectory. This value is computed by dividing the segment length by its duration. Assigning each segment to the corresponding speed category might, however, lead to a high degree of fragmentation. Consider the cases of slowing vehicles due to intersections, traffic lights, and obstacles. To avoid such fragmentation, a *sliding window* query is used and replaces the speed value of each segment by the mean value computed over a series of consecutive segments (Algorithm 2.4, Line 7). The splitting and classification of sub-trajectories is performed using these averaged values. The process is outlined in Algorithm 2.4. For each line segment l_i of a trajectory t, its mean speed is computed for a sliding

Algorithm 2.4. Segment trajectories according to speed profiles

Input: A set of trajectories T
Output: Sets of segmented trajectories C

```
1  begin
2      foreach t ∈ T do
3          foreach l_i ∈ t do
4              ▷ Average speed, ±w segments
5              v̄(l_i) ← Mean(v(l_{i−w}), ..., v(l_{i+w}))
6              k ← Category(v̄(l_i)) ▷ Determine category
7              C_k.add(l_i) ▷ Assign segment to respective category C_k
8          end
9      end
10 end
```

window of width $\pm w$ segments (Algorithm 2.4, Line 6). Then the segment is assigned to the corresponding category C_k according to the minimum and maximum speed of each category (cf. Lines 7 and 8 in Algorithm 2.4).

2.3.2 *Construction of Network Layers*

The next step is to use the trajectory portions that have been classified to construct the respective layers of the road network. This can be accomplished using an improved version of the *TraceBundle* algorithm.

This trace-based map construction algorithm employs heuristics to identify intersection nodes and "bundles" trajectories connected to them. Intersection nodes are clustered based on *changes in movement*. Such changes represent turns and are identified as changes in direction and speed. Clustering these turns based on (1) spatial proximity and (2) turn type results in *turn clusters*. Intersection nodes are then derived by clustering turn clusters based on proximity. Connecting the trajectories to intersection nodes and compacting them allows one to derive edges and consequently the entire geometry of the road network.

The clustering in the *TraceBundle* is based on two criteria, *proximity* and *angle difference*. *TraceBundle* allows only for one set of static parameters for both criteria. With different road types this often leads to erroneous clusters, e.g., generating multiple nodes for a single intersection, or generating a single node for multiple nearby intersections. To overcome this problem, a proximity-based expansion algorithm around turn samples based on turn similarity is used. The algorithm is detailed in Algorithm 2.5. It uses a segmented set of trajectories as input.

In a first step, all position samples are evaluated as to whether they represent turn samples based on a change of direction, adding them to the set of turn samples. The data recorded includes also the incoming and outgoing direction of the motion captured by the trajectory with respect to the specific turn sample (cf. Lines 10–16 in Algorithm 2.5).

Algorithm 2.5. Find intersections

Input: A set of trajectories T
Output: A set of Intersection nodes I

```
1  begin
2      S ← ∅ ▷ Turn samples
3      C ← ∅ ▷ Turn clusters
4      I ← ∅ ▷ Intersection nodes
5      α_max ▷ Angle difference threshold
6      d ▷ Proximity threshold
7      ▷ Position samples → turn samples
8      foreach t ∈ T do
9          foreach p_i ∈ t do
10             α_d ← AngularDiff(p_{i−1}, p_i, p_{i+1})
11             if α_d < α_max then
12                 α_in ← Angle(p_{i−1}, p_i) ▷ Incoming angle
13                 α_out ← Angle(p_i, p_{i+1}) ▷ Outgoing angle
14                 S.add(p_i, α_in, α_out)
15             end
16         end
17     end
18     ▷ Turn samples → turn clusters
19     foreach s ∈ S do
20         if s ∉ C.samples() then
21             ▷ Not yet considered
22             NN(p) ← S.findNN(s, d) ▷ Find neighboring, similar samples
23             c.add(ComputeTurnCluster(s, NN(p)))
24         end
25     end
26     ▷ Turn clusters → intersection nodes
27     foreach c ∈ C do
28         if c ∉ I.clusters() then
29             NN(c) ← C.findNN(c)
30             I.add(ComputeIntersections(c, NN_c))
31         end
32     end
33 end
```

This information is used to compute the directional similarity of turn samples. Samples that show a similar motion in terms of absolute direction and that are spatially close form turn clusters. The turn clusters are constructed bottom up by finding for each turn sample the set of nearest-neighbor samples considering also direction similarity. A threshold distance d (set to 25 m in the experiments) is used. Experiments showed that many turn clusters effectively have a radius much smaller than d, since turn samples of similar direction are either clustered together, or much farther away (relating to different intersections). The turn clustering approach is captured in Lines 20–23 of Algorithm 2.5.

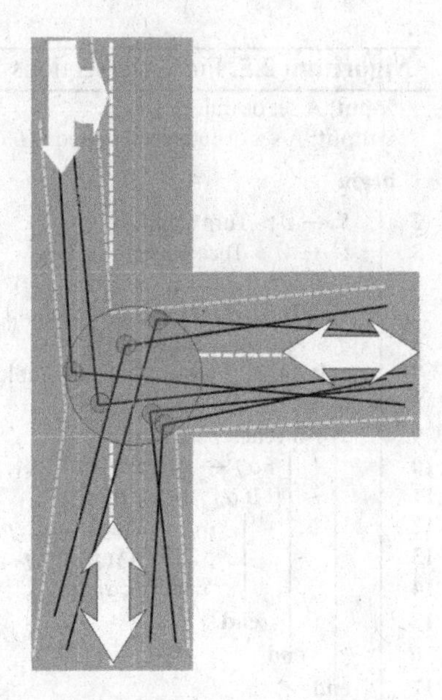

Fig. 2.8 Clustering of turn samples

Turn clusters stemming from different movement directions (left turn vs. right turn), but relating spatially to the same intersection, now need to be grouped together to produce one intersection node. Intersections are derived by scanning all turn clusters with respect to spatial coverage, i.e., smaller clusters if contained, will be absorbed by larger ones (cf. Lines 26–28 in Algorithm 2.5). Experiments showed this to be a very effective approach [5].

Figure 2.8 illustrates the outcome by contrasting an output from the *TraceBundle* algorithm with the current approach. Figure 2.9a shows how the clustering using static parameters in *TraceBundle* erroneously places nodes between actual intersections. Figure 2.9b shows the new approach with nodes being placed more accurately.

2.3.3 Conflation of Network Layers

The final step in *TraceConflation* is the fusion of the generated network layers into a single road network. The road network is constructed incrementally starting with higher speed layers and adding lower speed layers. The intuition for this is that higher speed layers correspond to major roads and they are generated more accurately, considering their more regular movement patterns and fewer GPS errors.

Fusing two network layers consists of (1) finding intersection node correspondences between the different network layers, (2) introducing new intersection nodes in existing edges of a higher layer, and (3) introducing new edges of lower layers for the uncommon portions of the road network.

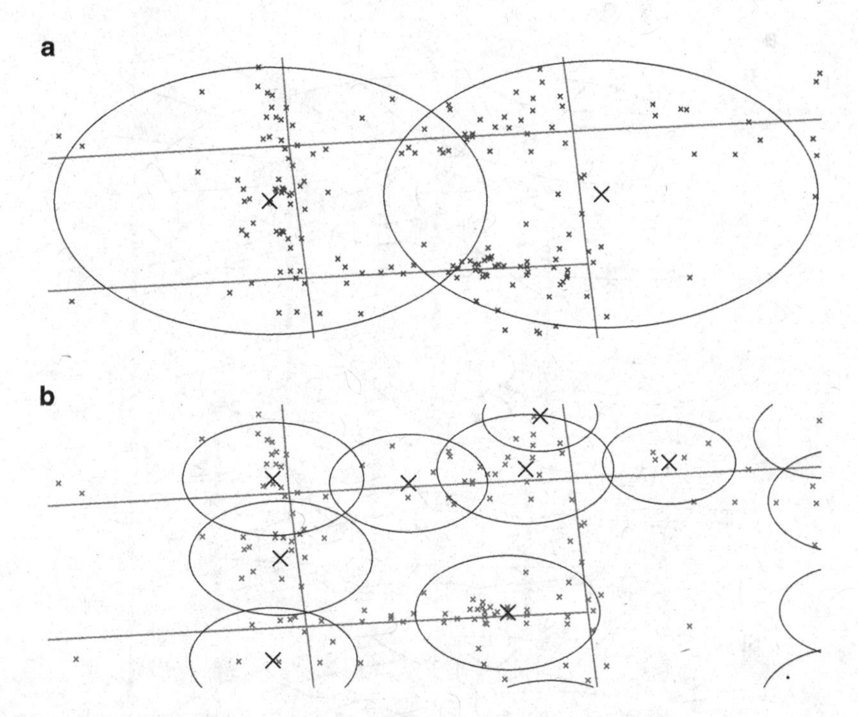

Fig. 2.9 Map construction—contrasting approaches. (**a**) Turn samples clustering in *TraceBundle*. (**b**) Turn samples clustering in *TraceConflation*

Figure 2.10 gives an example of this conflation process using the example of the Berlin dataset. Figure 2.10a shows the three maps that were constructed after segmenting the entire trajectory dataset of Fig. 2.1. Gray lines link the various connection points between the constructed networks. The final result is shown in Fig. 2.10b.

Starting with the fast and the medium network, common nodes are identified using spatial proximity. A threshold of 10 m was experimentally determined to be a good choice for determining if two nodes in the respective networks represent the same intersection [5].

The next step involves introducing new intersections in existing edges, e.g., in the fast network an edges exists, but the medium network has additional intersection nodes. Using a buffer region around intersection nodes of lower layers (e.g., medium) intersection nodes that are close to existing edges are identified. These new intersection nodes are then mapped to an existing edge to effectively split it.

Finally, new edges for uncommon portions of the layered network are added, e.g., an edge of the medium network missing in the fast network. Here, edges of lower layers are introduced by connecting them to previously introduced intersection nodes. Any intersection node that has not been introduced yet, since not connected

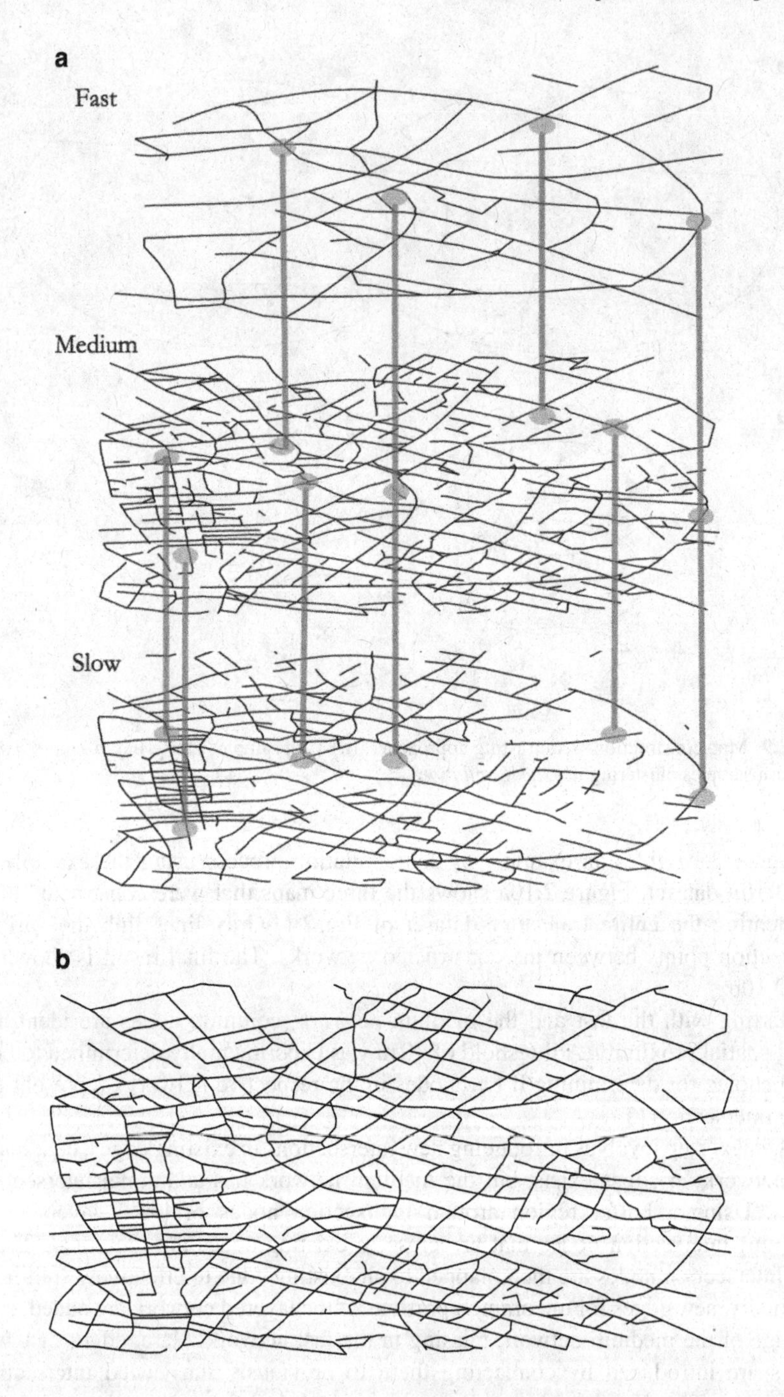

Fig. 2.10 Combining multiple network layers. (**a**) Fast-medium-slow network—*Berlin*. (**b**) Completely constructed map—*Berlin*

to the higher network, will be added as well. This accounts also for the case of adding complete (local) road network portions. A result of applying this conflation algorithm to road network layers is shown in Fig. 2.10.

2.4 Visual Summary

To simply showcase the result of the map construction algorithm in question, consider the example given in Fig. 2.11. Figure 2.11a shows the vehicle trajectories collected for *Berlin* in gray color (same figure as in Fig. 2.1). In Fig. 2.11b black lines indicate the constructed map and gray lines show the actual road network based on the OpenStreetMap dataset.

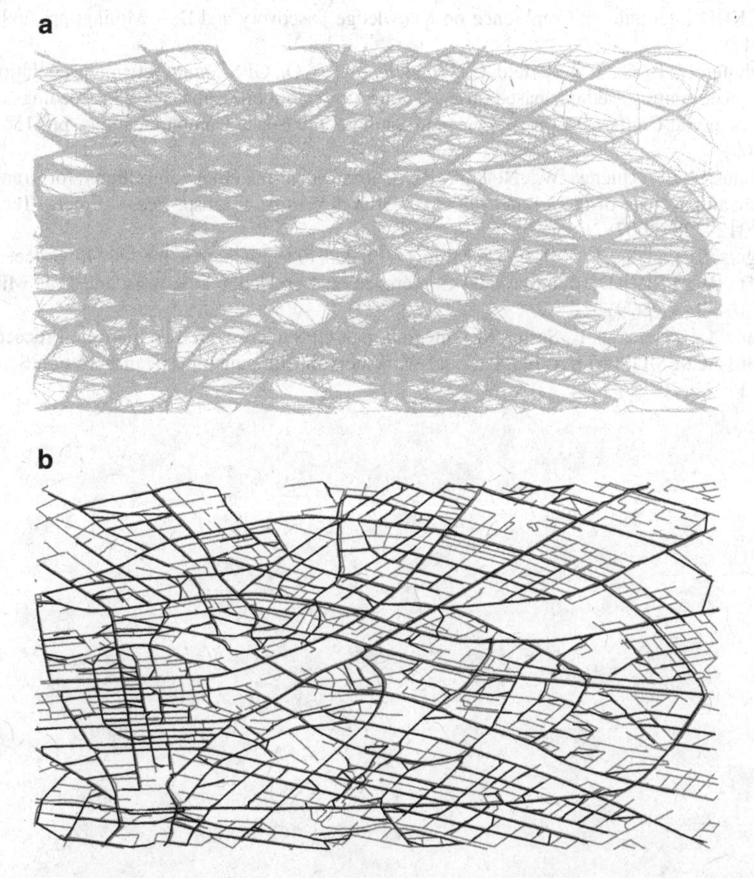

Fig. 2.11 Vehicle tracking data vs. constructed map overlaid on ground-truth OSM data. (**a**) Vehicle tracking data—*Berlin*. (**b**) Constructed map (in *black*) overlaid on ground-truth (in *gray*)

References

1. Bruntrup, R., Edelkamp, S., Jabbar, S., Scholz, B.: Incremental map generation with GPS traces. In: Proceedings of IEEE Intelligent Transportation Systems, pp. 574–579 (2005)
2. Cao, L., Krumm, J.: From GPS traces to a routable road map. In: Proceedings of 17th ACM SIGSPATIAL International Conference on Advances in Geographic Information Systems, pp. 3–12 (2009)
3. Fathi, A., Krumm, J.: Detecting road intersections from GPS traces. In: Proceedings of 6th International Conference on Geographic Information Science, pp. 56–69 (2010)
4. Karagiorgou, S., Pfoser, D.: On vehicle tracking data-based road network generation. In: Proceedings of 20th ACM SIGSPATIAL International Conference on Advances in Geographic Information Systems, pp. 89–98 (2012)
5. Karagiorgou, S., Pfoser, D., Skoutas, D.: Segmentation-based road network construction. In: Proceedings of 21st ACM SIGSPATIAL International Conference on Advances in Geographic Information Systems, pp. 450–453 (2013)
6. Liu, X., Biagioni, J., Eriksson, J., Wang, Y., Forman, G., Zhu, Y.: Mining large-scale, sparse GPS traces for map inference: comparison of approaches. In: Proceedings of 18th ACM SIGKDD International Conference on Knowledge Discovery and Data Mining, pp. 669–677 (2012)
7. Niehofer, B., Burda, R., Wietfeld, C., Bauer, F., Lueert, O.: GPS community map generation for enhanced routing methods based on trace-collection by mobile phones. In: Proceedings of 1st International Conference on Advances in Satellite and Space Communications, pp. 156–161 (2009)
8. Quddus, M., Ochieng, W., Noland, R.: Current map-matching algorithms for transport applications: state-of-the art and future research directions. Transp. Res. C Emerg. Technol. **15**, 312–328 (2007)
9. Rogers, S., Langley, P., Wilson, C.: Mining GPS data to augment road models. In: Proceedings of 5th ACM SIGKDD International Conference on Knowledge Discovery and Data Mining, pp. 104–113 (1999)
10. Zhang, L., Thiemann, F., Sester, M.: Integration of GPS traces with road map. In: Proceedings of 3rd ACM SIGSPATIAL International Workshop on Computational Transportation Science, pp. 17–22 (2010)

Chapter 3
Fréchet Distance-Based Map Construction Algorithm

Abstract This chapter presents an incremental track insertion algorithm for map construction that is based on partial map-matching of the trajectories to the graph. The Fréchet distance is used as part of the map-matching algorithm and to provide quality guarantees for the constructed map. One of the contributions of this work is to separate edge regions from vertex regions when providing quality guarantees, and to specify the extent of these regions. The algorithm itself is easy to implement and it can compute non-planar road configurations, such as bridges, if altitude information is provided with the trajectory measurements. While the description complexity of the computed road geometries can be guaranteed to be small when using a simplification algorithm, in practice the algorithm provides good results even without simplifying or merging trajectory portions that correspond to the same road segment.

3.1 Introduction

There are several different classes of algorithms to construct a road network from a set of input trajectories. The algorithm presented in this chapter is the incremental track insertion algorithm by Ahmed and Wenk [2], which incrementally adds one trajectory at a time to a road network. The advantage of using the continuous structure of the input trajectories is that the shape of the road segments (edges) in the reconstructed road network is preserved. The algorithm uses a new partial variant of map-matching based on the Fréchet distance, which is a distance measure suitable for comparing shapes of continuous curves. Quality guarantees for the constructed road network are provided in terms of the Fréchet distance, and in terms of edge regions and vertex regions. The output of the algorithm is an undirected embedded graph. If the input trajectories are embedded in the two-dimensional plane, then the output graph will be a planar embedded graph. But if the trajectories are embedded in three dimensions, i.e., with altitude information for each measurement, then the algorithm computes non-planar road configurations such as bridges.

The authors prove that there is a one-to-one correspondence with bounded complexity between *well-separated* portions of the original and the reconstructed edges. However, giving quality guarantees for portions of the graph where edges come close together, in particular in regions around vertices, is a much more challenging task. They provide the first attempt at reconstructing vertex regions and

M. Ahmed et al., *Map Construction Algorithms*, DOI 10.1007/978-3-319-25166-0_3

providing quality guarantees, without assuming very clean and extremely densely sampled data. They reconstruct intersections as sets of vertices within bounded regions (*vertex regions*), where the size of each set is bounded by the degree and the region is bounded by the minimum incident angle of the streets at that intersection. They can then guarantee that if the vertices are sufficiently far apart so that the vertex regions do not overlap, then the vertices of each region correspond to exactly one vertex in the original graph.

Map construction can be seen as a new type of geometric reconstruction problem in which the task is to extract the underlying geometric structure described by a set of movement-constrained trajectories, or in other words to reconstruct a geometric domain that has been sampled with continuous curves that are subject to noise. As a finite sequence of time-stamped position samples, each input trajectory represents a finite noisy sample of a continuous curve. While there are many map construction algorithms in the literature [5, 7, 10, 11, 14–17], most presented solutions are of a heuristic nature and do not provide quality guarantees. In the computational geometry community, however, there are several algorithms for the map construction problem that provide various kinds of quality guarantees. Chen et al. [9] reconstruct "good" portions of the edges (streets) and provide connectivities between these sections. They bound the complexity of the reconstructed graph and they guarantee a small directed Hausdorff distance between each original and corresponding reconstructed edge. Ge et al. [12] employ a topological approach by modeling the reconstructed graph as a Reeb graph. They define an intrinsic function which respects the shape of the simplicial complex of a given unorganized data point set, and they provide partial theoretical guarantees that there is a one-to-one correspondence between cycles in the original graph and the reconstructed graph. Aanjaneya et al. [1] view street networks as metric graphs and they prove that their reconstructed structure is homeomorphic to the original street network. Their main focus is on computing an almost isometric space with lower complexity, therefore they focus on computing the combinatorial structure but they do not compute an explicit embedding of the edges or vertices. Chazal et al. [8] also provide an algorithm for constructing metric graphs. They use a Reeb graph variant, the α-Reeb graph, and prove quality guarantees in terms of the Gromov-Hausdorff distance between the underlying metric graph and the constructed metric graph. The embedding is obtained from the Mapper algorithm for visualizing Reeb graphs [18]. All these map construction algorithms are point clustering approaches, in the sense that they consider the input as an unorganized set of measurement points and then they compute a neighborhood complex such as the Vietoris-Rips complex to serve as the input to their algorithms. The algorithm presented in this chapter is different in that it uses a track insertion approach but still provides quality guarantees.

3.2 Problem Statement and Data Model

Consider an underlying *original* graph (street network) $G_o = (V_o, E_o)$, modelled as an embedded undirected graph in $\mathbb{R}^2$ or $\mathbb{R}^3$. Let T be an input set of trajectories. Each trajectory is a sequence of measurements in $\mathbb{R}^2$ or in $\mathbb{R}^3$, and it is assumed to have sampled a connected sequence of edges in G_o (*street-path*). The error associated with each trajectory is modelled by a single precision parameter ε. Given an input set T of trajectories, which are assumed to be polygonal curves, and a precision parameter $\varepsilon > 0$, the *map construction task* is to compute an undirected *reconstructed* graph $G = (V, E)$ embedded in $\mathbb{R}^2$ or $\mathbb{R}^3$ that represents all curves in the set. The algorithm in this chapter guarantees that a *well-separated* portion of each edge in E_o corresponds to a sub-curve of an edge in E, and each vertex in V_o corresponds to a set of vertices in V.

V_o and V are assumed to be sets of vertices with degree > 2 and each edge is represented as a polygonal curve. It is also assumed that G_o is fully sampled, i.e., for each street $\gamma \in E_o$ there is a sampled sub-curve in the input curve set, T. Each edge of G_o is referred to as a *street*.

An input trajectory is a finite sequence of time-stamped position samples, which represents a finite noisy sample of a continuous curve. Generally, the measurements of the position samples are only accurate within certain bounds (measurement error), and the movement transition in between position samples can be modeled with varying accuracies depending on the application (sampling error). For this algorithm, the trajectories are modelled as piecewise linear curves, and of particular consideration will be trajectories of vehicles driving on a street network. In this case, the input curves in fact sample an $\omega/2$-fattening of a street-path, where ω is the street width. The r-fattening of a point set A is the Minkowski sum $A_r = A \oplus B(0, r)$, where $B(x, r)$ is the closed ball of radius r centered at x.

The single precision parameter ε captures the different kinds of noise as well as the street width. The algorithm uses the Fréchet distance [3] to measure the similarity between the shape of an input curve and a street-path. For two planar curves $f, g : [0, 1] \to \mathbb{R}^2$, the Fréchet distance δ_F is defined as

$$\delta_F(f, g) = \inf_{\alpha,\beta:[0,1]\to[0,1]} \max_{t\in[0,1]} \|f(\alpha(t)) - g(\beta(t))\|$$

where α, β range over continuous and non-decreasing reparametrizations, and $\|.\|$ denotes the Euclidean norm. By dropping the requirement on α and β to be non-decreasing, one obtains a distance measure that is called the *weak Fréchet distance*. An intuitive definition of the Fréchet distance is the minimum leash length that allows a man to walk on one curve and the dog on the other curve from the beginning of the curves to the end, both are allowed to control their speeds, but not allowed to backtrack.

To define the well-separability of streets, the following definition from [9] is used.

Definition 3.1 (α-Good). A point p on G is *α-good* if $B(p, \alpha) \cap G$ is a one-dimensional curve starting and ending at the boundary of $B(p, \alpha)$. A point p is *α-bad* if it is not α-good. A curve is α-good if all points on it are α-good.

3.3 Assumptions

The correctness of the algorithm depends on the following assumptions that are made about the original graph and the input data, see Figs. 3.1 and 3.2.

1. **Assumptions on G_o:**
   (a) Each street has a well-separated portion which is 3ε-good. This is referred to as a *good section.*
   (b) If for two streets γ_1, γ_2 there are points $p_1 \in \gamma_1$ and $p_2 \in \gamma_2$ with distance $\leq 3\varepsilon$, then γ_1 and γ_2 must share a vertex v_0, and the sub-curves $\gamma_1[p_1, v_0]$ and $\gamma_2[p_2, v_0]$ have Fréchet distance $\leq 3\varepsilon$ and they are fully contained in $B(v_0, 3\varepsilon/\sin\alpha)$. Here $\alpha = \angle p_1 v_0 p_2$, see Fig. 3.1b.

   From both assumptions follows that the minimum distance between two intersections is at least $3\varepsilon/\sin\alpha_1 + 3\varepsilon/\sin\alpha_2$, see Figs. 3.1a and 3.2. Assumption 1(a) states the minimum requirement to justify the existence of a street based on it being well-separated from other streets. Assumption 1(b) requires streets that are close together to converge to a vertex, and it discards streets that are close but do not share a vertex, because the input curves do not clearly distinguish between such streets. Note that the bound $3\varepsilon/\sin\alpha$ can be large for small values of α. So, the assumptions allow streets to be close together for a long time but restrict them to go far off once they are close to a vertex.
2. **Assumptions on the Input Data:**
   (a) Each input curve is within Fréchet distance $\varepsilon/2$ of a street-path in G_o.
   (b) All input curves sample an acyclic path in G_o.

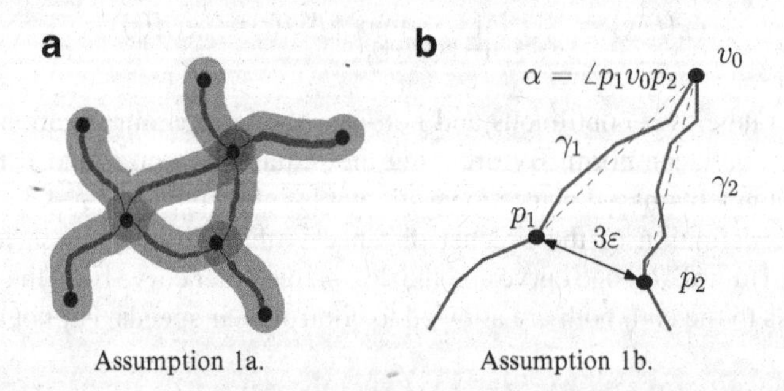

Fig. 3.1 Assumptions on the original graph. (**a**) Good sections on streets are marked as *bold*. (**b**) Visual illustration of Assumption 1b

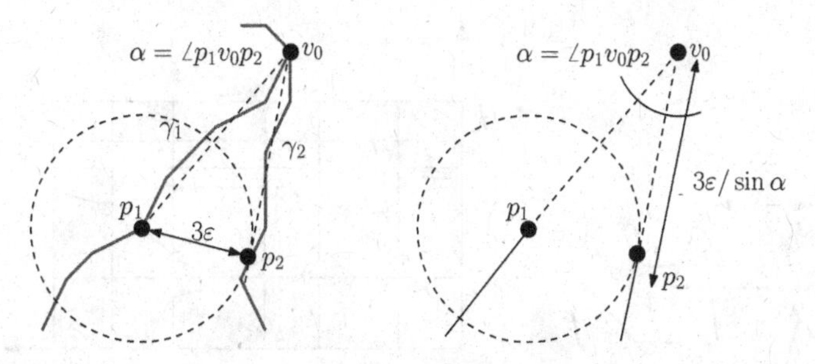

Fig. 3.2 Assumptions 1a and 1b imply minimum distance of $3\varepsilon/\sin\alpha$

Assumption 2(a) ensures that the street-path and the corresponding input curve have to be similar. Assumption 2(b) ensures that, during an incremental construction algorithm, the first curve in T that represents a new edge does not sample a cycle. The algorithm in fact only needs the unmatched portion (defined in Sect. 3.4) to sample an acyclic path. And even if it samples a cyclic path, such a cycle can be split in order to maintain this assumption.

Note that if a curve is well-sampled then the sampled curve is naturally within a bounded Fréchet distance of the original curve, which fulfills Assumption 2(a). In particular, for the test data of vehicle trajectories, the GPS device error is generally bounded. Also, data obtained from OpenStreetMap is generally sampled every second, and it captures every feature of the shape of the original street very well.

3.4 Free Space Diagrams and Map-Matching

The algorithm employs the concept of the *free space* F_ε and the *free space surface* FS_ε to identify clusters of sub-curves which sample the same street-path. For two planar curves $f, g : [0,1] \to \mathbb{R}^2$, and $\varepsilon > 0$, *free space* is defined as $F_\varepsilon(f,g) := \{(s_1, s_2) \in [0,1]^2 \mid \|f(s_1) - g(s_2)\| \leq \varepsilon\}$. The *free space diagram* $FD_\varepsilon(f,g)$ is the partition of $[0,1]^2$ into regions belonging or not belonging to $F_\varepsilon(f,g)$, see Fig. 3.3. In [3] it has been shown that $\delta_F(f,g) \leq \varepsilon$ if and only if there exists a curve within F_ε from the lower left corner to the upper right corner, which is monotone in both coordinates, see Fig. 3.3 for an illustration.

The *free space surface* $FS_\varepsilon(G,t)$ for a graph $G = (V,E)$ and a curve t (trajectory) is a collection of free space diagrams $FD_\varepsilon(e,t)$ for all $e \in E$ glued together according to the adjacency information of G, see Fig. 3.4. Note that the free space in $FS_\varepsilon(G,t)$ could be disconnected if the graph G is not connected.

The algorithm combines the idea of *map matching* and *partial curve matching* in order to map a curve partially to a graph. Buchin et al. [6] solved a *partial matching* for two curves by finding a monotone shortest path on a weighted FD_ε from lower left to upper right end point, where free space (white region) has weight 0 and non-free space (black region) has weight 1, see Fig. 3.5a. Alt et al. [4] solved the

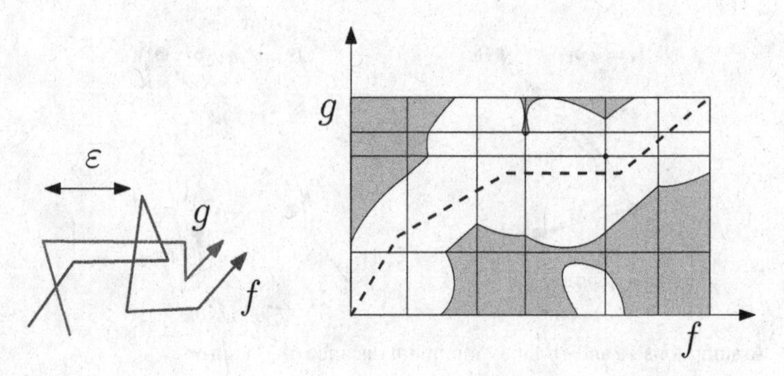

Fig. 3.3 FD_ε for two polygonal curves f, g. A monotone path is drawn in the free space

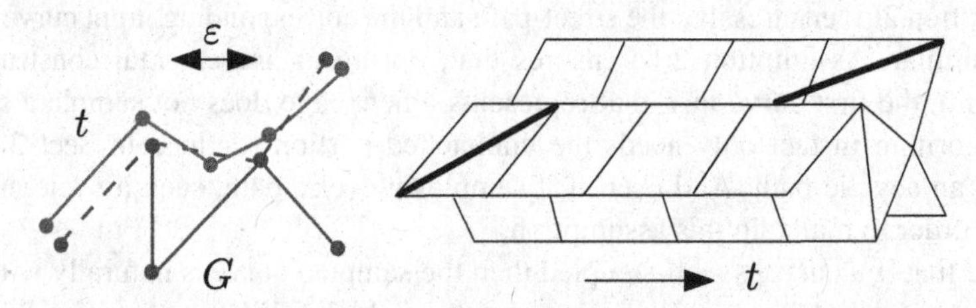

Fig. 3.4 FS_ε for a graph G and a curve t. An example path π is shown in *dashed* and a t-monotone path in FS_ε is highlighted in *bold*

map-matching problem of matching a curve t to a graph by finding a t-monotone path in the free space of FS_ε from any left to any right end point. Here, the path is not allowed to penetrate the black region. Note that a path in FS_ε corresponds to a path π in the graph, see Fig. 3.4. In the map construction setting one needs to find a t-monotone shortest path on the weighted free-space surface from any left end point to any right end point. However, finding such a shortest path on a weighted non-manifold surface is hard. Moreover, as the path can begin and end at any left or right end point, in some cases such a path does not provide the desired mapping, see Fig. 3.5a. The bold path is the desired one but the shortest path, with respect to Euclidean distance, is the dashed one.

3.5 Incremental Map Construction Algorithm

The incremental approach of this map construction algorithm is to insert one trajectory at a time into the graph, until all trajectories have been inserted. Each trajectory insertion step can be divided into two phases. The first phase (steps 1 and 2) computes a reconstructed graph, while the second phase (step 3) compresses the complexity of that graph. The simplicity of the algorithm relies on analyzing connectivity properties of the free space surface for the current trajectory and the current partially reconstructed graph, under the assumptions on the original graph and the input curves that are described in Sect. 3.3.

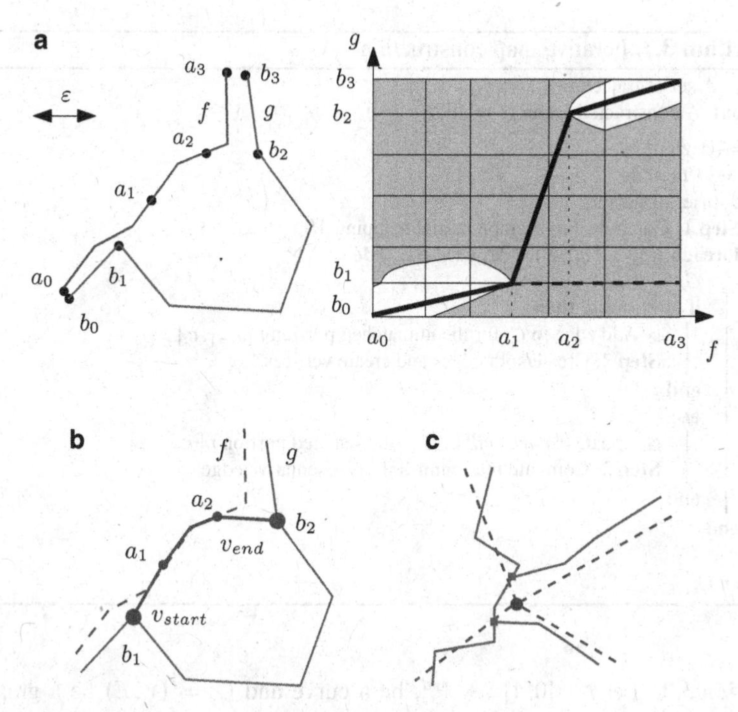

Fig. 3.5 Partially similar f and g. **(a)** $M_\varepsilon(f,g) = \{(f[a_0,a_1], g[b_0,b_1]), (f[a_1,a_2], null), (null, g[b_1,b_2]), (f[a_2,a_3], g[b_2,b_3])\}$. **(b)** Adding the unmatched portion of f as an edge. **(c)** G_o is shown in *dashed* and G in *bold*

Algorithm 3.1 shows the iterative map construction algorithm, where $G_i = (V_i, E_i)$ is the graph after the i-th iteration.

The total worst case runtime for adding the i-th trajectory $t_i \in T$ is $O(n_iN_{i-1} + k_c + (N_{i-1} + n_i)^2 \log(N_{i-1} + n_i)) = O((N_{i-1} + n_i)^2 \log(N_{i-1} + n_i))$, where N_{i-1} is the complexity of G_{i-1}, n_i is the complexity of t_i, and k_c is the number of created edges. A variant of the map construction algorithm runs only steps 1 and 2, which results in an improved runtime of $O(n_iN_{i-1} + k_c) = O(n_iN_{i-1})$. An implementation of this simplified variant is used in the experiments in Chap. 7. Step 3 lowers the complexity of the constructed edges by applying a minimum-link chain stabbing algorithm [13]. This ensures that the complexity of a constructed edge depends only on the complexity of the original street rather than on the complexity of the input curves.

Step 1: Compute Curve-Graph Partial Mapping $M_{1.5\varepsilon}(t_i, G_{i-1})$

The algorithm first breaks the trajectory into matched and unmatched portions by computing a *curve-graph partial mapping*.

Algorithm 3.1. Iterative map construction

Input: A set of trajectories $T = \{t_1, \ldots, t_m\}$
Output: A constructed graph $G = (V, E)$

1 $G_0 = (\emptyset, \emptyset)$
2 **for** $i \leftarrow 1$ *to* m **do**
3 ▷ Insert trajectory t_i
4 **Step 1:** Compute curve-graph partial mapping $M_{1.5\varepsilon}(t_i, G_{i-1})$
5 **foreach** $(t_i[a_{j-1}, a_j], X) \in M_{1.5\varepsilon}(t_i, G_{i-1})$ **do**
6 $G_i \leftarrow G_{i-1}$
7 **if** $X == null$ **then**
8 ▷ Add edge to G_i for the unmatched portion $t_i[a_{j-1}, a_j]$
9 **Step 2:** Create/split edges and create vertices
10 **end**
11 **else**
12 ▷ Update $e[start, end]$ in G_i with matched portion $t_i[a_{j-1}, a_j]$
13 **Step 3:** Compute minimum-link representative edge
14 **end**
15 **end**
16 **end**
17 **return** G_m

Definition 3.2. Let $t : [0, 1] \to \mathbb{R}^d$, be a curve and $G = (V, E)$ be a graph in $\mathbb{R}^d$, for $d = 2$ or $d = 3$. The *curve-graph partial mapping* $M_\varepsilon(t, G)$ consists of pairs $(t[a_{j-1}, a_j], X)$ where $0 = a_0 \leq a_1 \leq \ldots \leq a_k = 1$ such that every sub-curve $t[a_{j-1}, a_j]$ in a partition $t[a_0, a_1, \ldots, a_k]$ of t is either mapped to a portion $X = e[start, end]$ of an edge $e \in E$ with $\delta_F(t[a_{j-1}, a_j], e[start, end]) \leq \varepsilon$ or $X = null$ if no such edge exists. Here, edges are assumed to be polygonal curves parameterized over $[0, 1]$ and $0 \leq start \leq end \leq 1$. A sub-curve is called a *matched portion* if it has a non-null mapping, and an *unmatched portion* otherwise. We assume that all unmatched portions as well as all matched portions along the same edge are maximal.

At the beginning of this step, the algorithm computes a partial mapping $M_{1.5\varepsilon}(t_i, G_{i-1})$ which minimizes the total length of unmatched portions of t_i. For this, $FS_{1.5\varepsilon}(t_i, G_{i-1})$ is computed first, and then the white regions are projected onto the curve t_i. The union of all the white intervals on t_i, or equivalently on the parameter space $[0, 1]$, induces a partition $t_i[a_0, a_1, \ldots, a_{k_u+k_c}]$ of t_i into maximal *white intervals* and *black intervals*; the latter fill the gap between non-overlapping white intervals. This defines matched portions $(t_i[a_{j-1}, a_j], e[start, end])$ for white intervals, and unmatched portions $(t_i[a_{j-1}, a_j], null)$ for black intervals.

Note that the objective for the partial mapping to minimize the total length of unmatched portions of t_i is one-sided. This corresponds to computing a shortest monotone path in $FS_{1.5\varepsilon}(t_i, G_{i-1})$ that minimizes the total length of the unmatched portions along the curve t_i. Such a desired shortest path in the free space surface is therefore allowed to go through the black region along the graph direction without adding any additional cost. Hence, the length of the path is measured *only within the unmatched portion and only along the curve direction.*

The projection approach to consider white and black intervals only is justified by Assumption 1(a) and (b) on the street network, which ensure that no two streets are closer than 3ε to each other or otherwise they would have to share a vertex. This implies that for each $t_i[a_{j-1}, a_j]$ that samples a portion of a good section that no more than one $e[start, end]$ can exist, and hence $e[start, end]$ is uniquely defined for each matched portion.

This step outputs the curve-graph partial mapping as a list of black and white interval/edge pairs, ordered by the interval start points on t_i. Such a list can be computed in $O(n_i N_{i-1})$ total time.

Step 2: Create/Split Edges and Create Vertices

In this step of the algorithm a new edge is created, in constant time, for each unmatched portion $(t_i[a_{j-1}, a_j], null) \in M_{1.5\varepsilon}(t_i, G_{i-1})$. By construction, the previous and next intervals of a black interval are either *null* (the interval is the first or last element in the list) or white. Assume them to be non-*null*, because only in this case a vertex has to be created. Let $(t_i[a_{j-2}, a_{j-1}], e_{prev}[start, end])$ be the mapped pair for the previous interval, and let $(t_i[a_j, a_{j+1}], e_{next}[start', end'])$ be the pair for the next interval. Then $u = e_{prev}[end]$ and $v = e_{next}[start']$ are points on the polygonal curves e_{prev} and e_{next}, respectively. If $start > 0$ then u is inserted into V_i as a new vertex, and if $end' < 1$ then v is inserted as a new vertex. Otherwise they already exist as vertices in V_i. Note that multiple vertices might be created for a single vertex in the original graph, see Fig. 3.5c. Then the three steps below are executed to split the existing edges and to create a new edge for the unmatched trajectory portion.

1. Split e_{prev} into $e_{prev}[0, end]$ (ending in u) and $e_{prev}[end, 1]$ (starting at u).
2. Split e_{next} into $e_{next}[0, start']$ (ending in v) and $e_{next}[start', 1]$ (starting at v).
3. Insert $u \circ t_i[a_{j-1}, a_j] \circ v$ as an edge into E_i, where $\circ$ denotes concatenation. Hence, the unmatched trajectory portion is connected to G_i at vertices u and v.

For example, in Fig. 3.5a, consider g as an edge in G_{i-1} and f as t_i. Figure 3.5b shows the addition of the unmatched portion of f as an edge in the graph; here $e_{prev} = e_{next}$.

Step 3: Compute Minimum-Link Representative Edge

After finishing steps 1 and 2, a graph has successfully been constructed for trajectories $t_1, \ldots, t_i$, which could be considered the output graph already. The purpose of this step is to reduce the complexity of the matched portions by computing a minimum-link representation for each $(t_i[a_{j-1}, a_j], X) \in M_{1.5\varepsilon}(t_i, G_{i-1})$, where $X \neq null$.

The problem statement is as follows: Given two polygonal curves f, g which both have bounded Fréchet distance to another polygonal curve γ such that $\delta_F(\gamma, f) \leq \varepsilon$ and $\delta_F(\gamma, g) \leq \varepsilon/2$, the task is to find a minimum-link representative curve e for γ such that $\delta_F(\gamma, e) \leq \varepsilon$. This is accomplished by first constructing a combined vertex-sequence $v_1, \ldots, v_s$ of vertices of f and g by following a monotone path in the free space of $FD_{1.5\varepsilon}(f, g)$. Let γ' be the polygonal curve associated with this sequence. According to Lemma 3.1 below, it holds that $\delta_F(\gamma, \gamma') \leq 2\varepsilon$. Then the minimum-link algorithm by Guibas et al. [13] is run on the sequence of balls $B(v_1, 2\varepsilon), \ldots, B(v_s, 2\varepsilon)$, with the additional restriction that the vertices of the output curve must lie within $g^{\varepsilon/2}$. While the resulting path e obtained by this modified algorithm might not be the minimum-link path for this particular sequence of balls, the complexity of the resulting polygonal curve is at most the complexity of the original curve γ, since $(\gamma')^{2\varepsilon}, f^{\varepsilon}$ and $g^{\varepsilon/2}$ all contain γ. Using the triangle inequality it can be proven that $\delta_F(e, \gamma) \leq \varepsilon$.

In step 3 of the algorithm, the edges of G_i are updated for each white interval with the minimum-link curve computed by the algorithm described above, where the polygonal curve f corresponds to $e[start, end]$ and g corresponds to $t_i[a_{j-1}, a_j]$. The time complexity to compute such a path is $O(n^2 \log n)$, where n is the number of vertices in γ'. The total runtime to update all edges when inserting the i-th trajectory is $O((N_{i-1} + n_i)^2 \log(N_{i-1} + n_i))$.

Lemma 3.1. *Let f and g be curves that sample a curve γ such that $\delta_F(f, \gamma) \leq \varepsilon$ and $\delta_F(g, \gamma) \leq \varepsilon/2$. Then any curve γ' comprised of vertices of f and g based on their order of a monotone path in the free space of $FD_{1.5\varepsilon}(f, g)$ has $\delta_F(\gamma, \gamma') \leq 2\varepsilon$.*

3.6 Quality Analysis

Lemma 3.2 in this section states that if an input curve $t \in T$ samples a good section of a street or a street-path, then that street-path is unique in the original graph G_o. It can be proven using a loop invariant that, if every edge in a path has exactly one good section, then after adding the i-th curve t_i, the reconstructed graph G_i preserves all paths of G_o^i. Here, *preserving* a path means that all good sections have Fréchet distance at most ε to the original street, and all vertices lie in the vertex region around the original vertices. And G_o^i is the sub-graph of G_o fully sampled by $t_1, \ldots, t_i$.

Lemma 3.2. *For each $t \in T$ there exists a mapping $M_\delta(t, G_o) = \{(t[0, a_1], \gamma_1[b_0, 1]), (t[a_1, a_2], \gamma_2[0, 1]), \ldots (t[a_{k-1}, 1], \gamma_k[0, b_k])\}$ for $\varepsilon/2 \leq \delta < 2.5\varepsilon$. And for $k \geq 3$ if t samples a good section of γ_1 and γ_k then $\gamma_1 \circ \gamma_2 \circ \gamma_3 \circ \ldots \circ \gamma_k$ is unique, otherwise $\gamma_2 \circ \gamma_3 \circ \ldots \circ \gamma_{k-1}$ is unique.*

3.6.1 Recovering Good Sections

Lemma 3.3 below provides a quality guarantee for good sections of streets. If a street $\gamma \in E_o$ is sampled by a set of input curves $T_\gamma = \{t_1, t_2, \ldots, t_k\}$, then it is shown that the map construction algorithm presented in this chapter reconstructs each good section of γ as one edge $e \in E$.

Lemma 3.3. *For each edge $\gamma \in E_o$, if β is a good section of γ and there exists a non-empty set of input curves $T_\gamma = \{t_1, t_2, \ldots, t_k\} \subseteq T$ such that for every $t_i \in T_\gamma$, $\delta_F(t_i[start_i, end_i], \gamma) \leq \varepsilon/2$, then there exists only one $e \in E$ such that $\delta_F(e[start, end], \beta) \leq \varepsilon$ and the complexity of $e[start, end]$ is at most the complexity of β.*

Proof. The proof follows from the construction of the curve-graph partial mapping $M_{1.5\varepsilon}(t_i, G_{i-1})$. When the first curve of the set is added to the graph, it is identified as an unmatched portion. And after that, all other sub-curves that sample γ will be identified as matched portions. Once the first matched portion is processed (i.e., for the second curve in T_γ), the minimum-link curve is computed which ensures the minimum complexity and which has Fréchet distance at most ε from the original street. Thus, all the other sub-curves will also appear as matched portions in $M_{1.5\varepsilon}(t_i, G_{i-1})$, that means for all $t_i \in T_\gamma$ that sample β only one edge will be created in the graph. □

3.6.2 Bounding Vertex Regions

The following provides a description of the vertex region, R_{v_0}, for reconstructed vertices around the original vertex, v_0. The analysis is performed for a three-way intersection, but it can easily be extended to an arbitrary n-way intersection.

Consider three streets γ_1, γ_2 and γ_3 incident to a vertex v_0. Let v_1^3 and v_3^1 be the farthest pair of points from v_0 on γ_1 and γ_3 such that $d(v_3^1, v_1^3) \leq 3\varepsilon$, see Fig. 3.6a. Here, $\angle v_3^1 v_0 v_1^3 = \alpha_3$, and according to Assumption 1(b), the line segments $v_0 v_1^3$ and $v_0 v_3^1$ are fully contained in $B(v_0, 3\varepsilon/\sin\alpha_3)$.

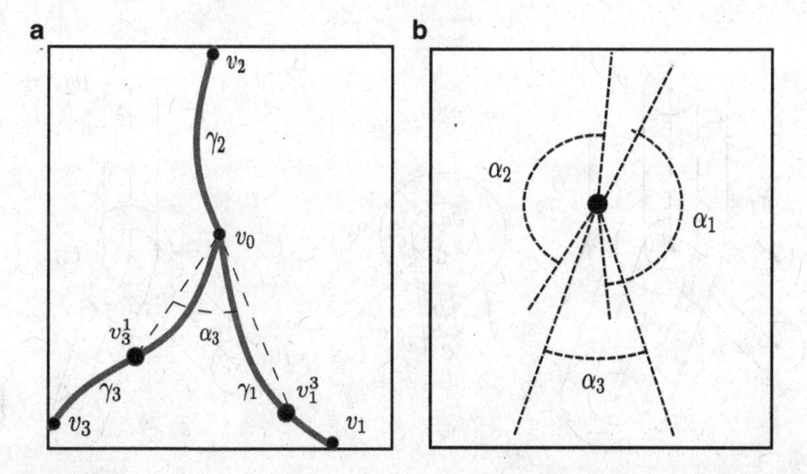

Fig. 3.6 A three-way intersection. **(a)** Three streets $\gamma_1, \gamma_2, \gamma_3$ incident to vertex v_0. **(b)** Different angles $\alpha_1, \alpha_2, \alpha_3$

For a three-way intersection there can be several input curves that sample three different street-paths going through v_0, and in order to reconstruct the vertex only two of them that sample two different street-paths are needed. Depending on which two input curves are used in which order, the reconstructed vertex can be in a different location in the vertex region. Note that although the constructed map is undirected, the input curves are sequences of samples and thus have directions. Therefore each street-path can be sampled by two input curves with opposite directions. For analysis purposes, a minimal collection of sets of input trajectories is defined, such that each set describes a region, and the union of these region forms R_{v_0}.

The first set T_1 contains trajectories t_i and t_j for $i < j$, where t_i samples street-path $\pi_1 = \gamma_1 \circ \gamma_2$ and t_j samples $\pi_2 = \gamma_2 \circ \gamma_3$. From Assumption 2(a) follows that $\delta_F(t_i, \pi_1) \leq \varepsilon/2$ and $\delta_F(t_j, \pi_2) \leq \varepsilon/2$. In the i-th iteration, t_i is inserted into G_{i-1} as an edge e, and in the j-th iteration t_j is considered. To compute $M_{1.5\varepsilon}(t_j, G_j)$, the intersection of $t_j^{1.5\varepsilon}$ and e is computed and a vertex at the intersection point is created which defines the mapping of a partition point of a matched and an unmatched portion of t_j. By construction, e could lie anywhere within π_1^{ε} and t_j could lie anywhere within $\pi_2^{\varepsilon/2}$. The intersection region that contains all possible intersections is $\pi_1^{\varepsilon} \cap \pi_2^{2\varepsilon}$. Since vertices are created only on the boundaries of the intersection region, no vertices are created if the street-paths are closer than 1.5ε from each other, so the region defined by T_1 is $R_{v_0}(T_1) = (\pi_1^{\varepsilon} \cap \pi_2^{2\varepsilon}) \setminus (\pi_1^{\varepsilon} \cap \pi_2^{\varepsilon})$, see Fig. 3.7a. Each such set contains two input trajectories, and for a three-way vertex there are six sets, $T_1, \ldots, T_6$. The vertex region R_{v_0} is then the union $R_{v_0} = R_{v_0}(T_1) \cup \ldots \cup R_{v_0}(T_6)$, see Fig. 3.7b for an example.

The above analysis can be extended to an arbitrary λ-way intersection, where there are $\lambda(\lambda-1)/2$ possible paths and $r = (\lambda(\lambda-1)/2)!/(\lambda(\lambda-1)/2-(\lambda-1))!$ sets of input curves are involved. The vertex region is then defined as $R_{v_0} = R_{v_0}(T_1) \cup \ldots \cup R_{v_0}(T_r)$.

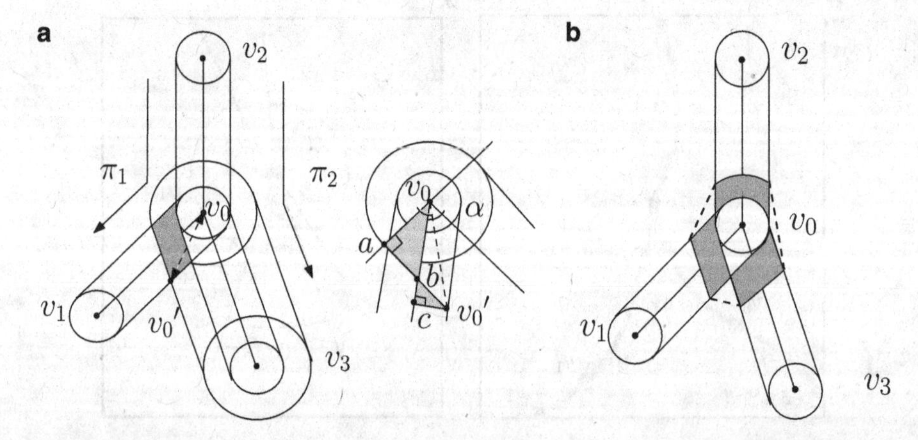

Fig. 3.7 The vertex region of a three-way intersection. (**a**) Three way intersection region considering only T_1. (**b**) The *shaded* region is $R(v_0)$

Lemma 3.4. *If $v_0' \in R(v_0)$ is the farthest point from a vertex $v_0 \in V_o$ with degree λ, then $\varepsilon \leq d(v_0, v_0') \leq \varepsilon / \sin\alpha\sqrt{5 + 4\cos\alpha}$, where $\alpha = \min_i \alpha_i$.*

Proof. In Fig. 3.7a, when considering the triangle Δv_0ab one obtains $\overline{v_0b} = \overline{v_0a}/\cos(\pi/2 - \alpha) = 2\varepsilon/\sin\alpha$. And when considering $\Delta v_0'bc$ one obtains $\overline{bc} = \overline{cv_0'}/\tan\alpha = \varepsilon/\tan\alpha$. From this follows that $d(v, v_0) = \sqrt{(\overline{v_0c})^2 + (\overline{v_0'c})^2} = \sqrt{\varepsilon^2 + ((2\varepsilon + \varepsilon\cos\alpha)/\sin\alpha)^2} = \sqrt{\varepsilon^2/\sin\alpha^2(5 + 4\cos\alpha)} = \varepsilon/\sin\alpha\sqrt{5 + 4\cos\alpha}$. □

3.7 Visual Summary

Figure 3.8 shows an example output of the map construction algorithm from this chapter for vehicle trajectories collected for *Berlin*.

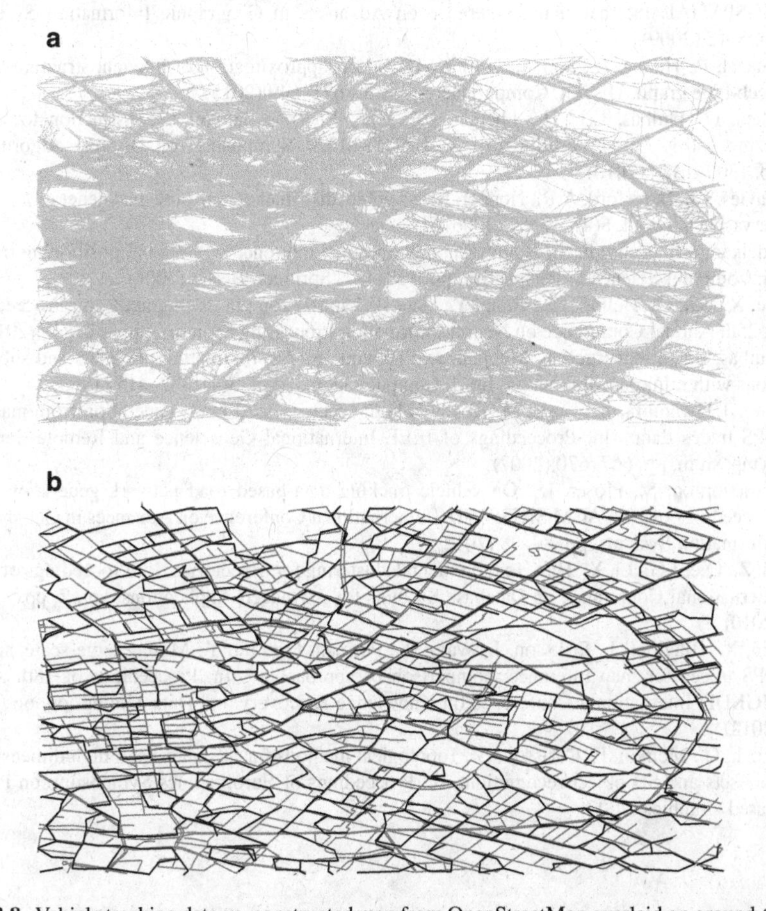

Fig. 3.8 Vehicle tracking data vs. constructed map from OpenStreetMap overlaid on ground-truth. (**a**) Vehicle tracking data—*Berlin*. (**b**) Constructed map (in *black*) overlaid on ground-truth map (in *gray*)

References

1. Aanjaneya, M., Chazal, F., Chen, D., Glisse, M., Guibas, L.J., Morozov, D.: Metric graph reconstruction from noisy data. In: Proceedings of 27th ACM Symposium on Computational Geometry, pp. 37–46 (2011)
2. Ahmed, M., Wenk, C.: Constructing street networks from GPS trajectories. In: Proceedings of 20th Annual European Symposium on Algorithms, pp. 60–71 (2012)
3. Alt, H., Godau, M.: Computing the Fréchet distance between two polygonal curves. Int. J. Comput. Geom. Appl. **5**, 75–91 (1995)
4. Alt, H., Efrat, A., Rote, G., Wenk, C.: Matching planar maps. J. Algorithms **49**, 262–283 (2003)
5. Bruntrup, R., Edelkamp, S., Jabbar, S., Scholz, B.: Incremental map generation with GPS traces. In: Proceedings of IEEE Intelligent Transportation Systems, pp. 574–579 (2005)
6. Buchin, K., Buchin, M., Wang, Y.: Exact algorithm for partial curve matching via the Fréchet distance. In: Proceedings of ACM-SIAM Symposium on Discrete Algorithms (SODA09), pp. 645–654 (2009)
7. Cao, L., Krumm, J.: From GPS traces to a routable road map. In: Proceedings of 17th ACM SIGSPATIAL International Conference on Advances in Geographic Information Systems, pp. 3–12 (2009)
8. Chazal, F., Huang, R., Sun, J.: Gromov-Hausdorff approximation of filament structure using Reeb-type graph. Discret. Comput. Geom. **53**(3), 621–649 (2015)
9. Chen, D., Guibas, L.J., Hershberger, J.E., Sun, J.: Road network reconstruction for organizing paths. In: Proceedings of 21st ACM-SIAM Symposium on Discrete Algorithms, pp. 1309–1320 (2010)
10. Davies, J.J., Beresford, A.R., Hopper, A.: Scalable, distributed, real-time map generation. IEEE Pervasive Comput. **5**(4), 47–54 (2006)
11. Edelkamp, S., Schrödl, S.: Route planning and map inference with global positioning traces. In: Computer Science in Perspective, pp. 128–151. Springer, Berlin (2003)
12. Ge, X., Safa, I., Belkin, M., Wang, Y.: Data skeletonization via Reeb graphs. In: Proceedings of 25th Annual Conference on Neural Information Processing Systems, pp. 837–845 (2011)
13. Guibas, L., Hershberger, J., Mitchell, J., Snoeyink, J.: Approximating polygons and subdivisions with minimum-link paths. Int. J. Comput. Geom. Appl. **3**(4), 383–415 (1993)
14. Guo, T., Iwamura, K., Koga, M.: Towards high accuracy road maps generation from massive GPS traces data. In: Proceedings of IEEE International Geoscience and Remote Sensing Symposium, pp. 667–670 (2007)
15. Karagiorgou, S., Pfoser, D.: On vehicle tracking data-based road network generation. In: Proceedings of 20th ACM SIGSPATIAL International Conference on Advances in Geographic Information Systems, pp. 89–98 (2012)
16. Li, Z., Lee, J.G., Li, X., Han, J.: Incremental clustering for trajectories. In: Proceedings of 15th International Conference on Database Systems for Advanced Applications, vol. 2, pp. 32–46 (2010)
17. Liu, X., Biagioni, J., Eriksson, J., Wang, Y., Forman, G., Zhu, Y.: Mining large-scale, sparse GPS traces for map inference: comparison of approaches. In: Proceedings of 18th ACM SIGKDD International Conference on Knowledge Discovery and Data Mining, pp. 669–677 (2012)
18. Singh, G., Mémoli, F., Carlsson, G.: Topological methods for the analysis of high dimensional data sets and 3D object recognition. In: Proceedings of Eurographics Symposium on Point-Based Graphics (2007)

Chapter 4
Density-Based Map Construction Pipeline

Abstract This chapter presents a density-based algorithm pipeline to construct a road map from a set of input trajectories. In the first step of the pipeline a data density function is computed and an undirected skeleton graph is constructed using grayscale skeletonization of the density. In several different steps the pipeline then uses the trajectory data to refine the topology and the geometry of the graph using the continuous information represented in the trajectories. Additional information including edge directions, lanes, and turn lanes are also added to the graph in these later steps.

4.1 Introduction

As discussed in Chap. 1, there are several different output models and several different map construction approaches in the literature. This chapter presents the map construction pipeline by Biagioni and Eriksson [2], which is a density-based point clustering algorithm with several unique features. The authors have made the implementation of their algorithm (all steps except the last one) publicly available, see for example http://www.mapconstruction.org/. The first step is to compute a density function from a point representation of the input trajectories, similar to a kernel density estimate. From this density, a *skeleton* is computed that is modelled as an undirected planar graph. This approach is common in density-based map construction algorithms [3, 4, 11], and such a skeleton could be interpreted as a valid output of a map construction algorithm already. The remaining steps of the pipeline then use the continuity of the trajectories to clean up various kinds of noise and to add additional information to the graph such as edge directions, lanes, and turn lanes.

The algorithm pipeline thus exemplifies the use of (1) density-based map construction algorithms, (2) gray-scale skeletonization akin to image processing approaches, (3) different map construction output models, and (4) the combination of point-wise methods with trajectory-based methods to infer more detailed topological and directional map information.

M. Ahmed et al., *Map Construction Algorithms*, DOI 10.1007/978-3-319-25166-0_4

(1) **Density estimation**: Use piecewise linearly interpolated input trajectories to compute a density over the planar domain.
(2) **Skeleton computation**: Compute an undirected embedded *skeleton* graph from the density. Edges are associated with their pixel-values and densities.
(3) **Topology refinement**:

a. **Density-aware map-matching**: Map-match the input trajectories onto the skeleton, using the density values on the graph edges to guide the map-matching. This results in a directed embedded graph, keeping only the directions represented in the trajectory edge traversals.
b. **Edge pruning**: Discard directed edges with at most one traversed trajectory.
c. **Vertex merging**: Merge intersection vertices that are too close together, if this does not decrease the number of well-matched trajectories.
d. **Traversal-aware map-matching**: Map-match the trajectories onto the current graph, using the number of previous trajectory traversals to guide the map-matching.
e. **Edge pruning**: Again, discard directed edges with at most one traversed trajectories. This prunes spurious spurs that remained after breaking spurious cycles in the first pruning step.

(4) **Geometry refinement**:

a. **K-means refinement**: Uniformly sample the current graph embedding, interpreting these samples as means, and run the k-means algorithm with the input measurement data to adjust the location of the samples and hence the embedding of the graph.
b. **Turn lanes**: Compute a more accurate representation of the observed traffic flow at each intersection, by performing k-means refinement on uniformly sampled pairs of edge transitions around each intersection. This results in directed turn lanes and straight lanes at each intersection.

Fig. 4.1 Steps of the map construction pipeline

4.2 Map Construction Pipeline

For the map construction task a set of *trajectories* is given as input, and the goal is to compute a *street map* that represents all trajectories. Each trajectory is a sequence of *measurements*, where each measurement consists of a point (latitude/longitude or (x, y)-coordinate after suitable projection), a time stamp, and optionally additional information such as vehicle heading or speed. A street map is generally modelled as some sort of an embedded graph. The presented map construction pipeline generally works in the geographic coordinate system using latitude/longitude as coordinates, but during some steps it uses a grid-mapping into Euclidean space.

The map construction pipeline by Biagioni and Eriksson [2] proceeds in four steps. See Fig. 4.1 for a summary. The first two steps, **(1) Density estimation** and **(2) Skeleton computation**, can be seen together as a density-based map construction algorithm on its own that outputs an undirected planar geometric graph as the street map. This undirected graph is referred to as the *skeleton*. Each trajectory is represented as a sequence of latitude/longitude measurements that are interpolated with "straight line" segments. For the density estimation step, the bounding box of all input measurements is computed in geographic coordinates, and is then mapped to Euclidean coordinates using an equirectangular projection. (This may also be referred to as "unprojected".) The density is then computed in Euclidean coordinates from a histogram on a $1 \times 1\,\text{m}^2$ pixel grid, considering all points along the polygonal lines described by the trajectories. Then a multi-level grayscale skeletonization algorithm is performed on this density map to obtain a set of skeleton pixels, and finally an undirected embedded planar graph is extracted to form the *skeleton graph*. These two steps are explained in more detail in Sect. 4.3.

In the remaining two steps, **(3) Topology refinement** and **(4) Geometry refinement**, several passes are made over the skeleton graph to reduce topological and geometric noise, and to add lanes, lane directions, and turn lanes. Here, the continuous trajectory information is used to infer valid (directed) transitions within the skeleton graph. A hidden Markov model-based map-matching algorithm is used to map the trajectories onto the skeleton in order to perform these refinements. In the geometry refinement step the polygonal shape of each edge is adjusted using a k-means clustering algorithm, and turn lanes are added at each intersection to represent valid transitions as well as the geometric shapes of the underlying turn lanes. These two steps are explained in more detail in Sect. 4.4. The final graph is an embedded directed graph with multiple lanes as well as a traffic flow-based intersection geometry with turn lanes. Figure 4.2 illustrates the steps of the pipeline on an example.

4.3 Density Estimation and Skeleton Computation

The first two steps of the pipeline, **(1) Density estimation** and **(2) Skeleton computation**, take as input a set of trajectories, and produce an undirected skeleton graph as the output, representing an initial draft of the underlying road network that has been travelled by the trajectories. After computing the bounding box of all input measurements, the latitude/longitude geographic coordinates are trivially mapped ("unprojected") into the Euclidean plane. Then a histogram is computed on a $1 \times 1\,\text{m}^2$ pixel grid, counting the number of trajectories that pass through each pixel. For this, the trajectories are piecewise linearly interpolated between the measurements. In order to account for noise and road width, the histogram is then convolved with a Gaussian kernel with $\sigma = 8.5\,\text{m}$. This smoothed histogram is an approximation of a kernel density estimate. In the literature, several different approaches have been used to generate densities from the trajectory data [3, 4, 11, 12]. And any other desired density estimation algorithm could be used for this step.

For a given density function over the bounding box domain, the goal of the second step is to compute a skeleton graph by first identifying skeleton pixels

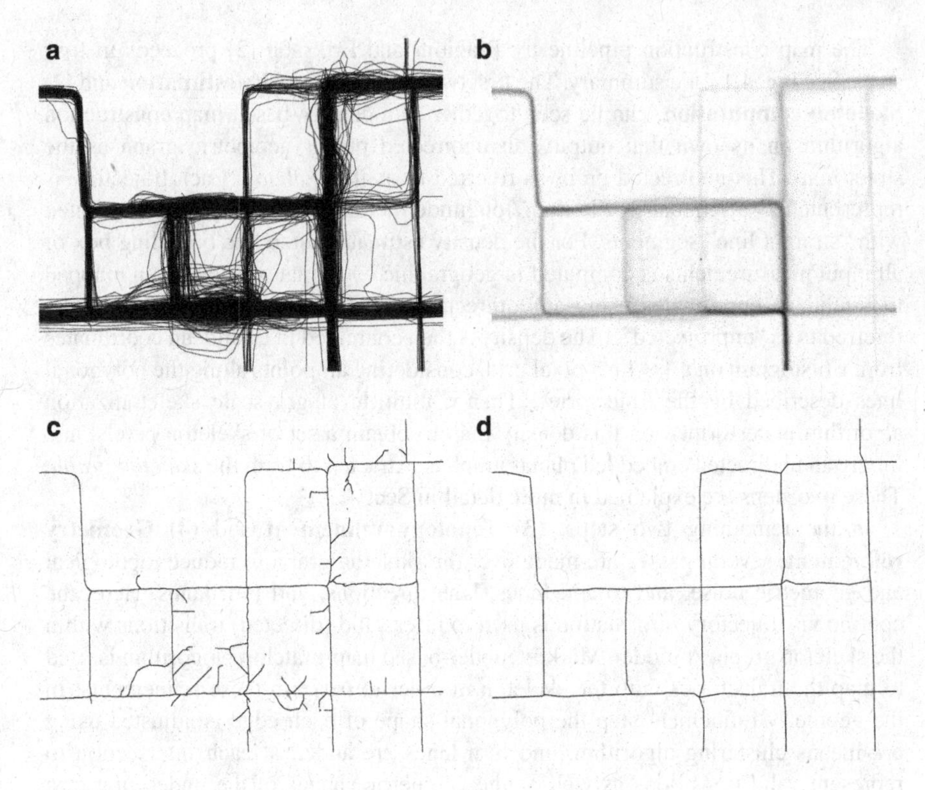

Fig. 4.2 An example illustrating the steps of the map construction pipeline. (**a**) Raw data. (**b**) Density. (**c**) Skeleton. (**d**) Topology-refined graph

and later inferring the adjacency structure between vertices. Identifying skeleton pixels is in principle a grayscale skeletonization task. Skeletonization, or thinning, of both binary and grayscale images has been extensively studied in the literature, see [8, 9, 14] and see [7, 10] for surveys. Algorithms for binary thinning include erosion-based approaches, medial axis computation, or topological approaches [5]. As part of the map construction pipeline, Biagioni and Eriksson describe a multi-pass thresholding algorithm that uses the binary thinning algorithm by Zhang and Suen [15] for a decreasing sequence of thresholds. The thinning algorithm itself is an iterative erosion algorithm based on local pixel neighborhoods. Overall, the multi-pass algorithm results in a set of pixels gradually added to the skeleton, in perceived order of higher to lower importance. The *gray-scale skeleton*, an image representation of the skeleton pixels with intensities taken from the computed trajectory density function, is used in step (3) below.

Finally, the set of skeleton pixels are used as input for a graph extraction module. Following the "combustion" technique from [11], pixels are classified into intersection pixels based on local neighborhood patterns of pixels. Then these pixels are used as seeds for multiple traversals to identify connected components of intersection pixels, and of edge pixels, and each such connected component describes a vertex or an edge, respectively. The embedding for the vertices is computed as the average location of the vertex-pixels. For each edge, the traversal already generates a sequence of edge-pixels, which can be interpreted as a very dense polygonal curve yielding an embedding for the edge. This curve is, however, simplified to a polygonal curve with fewer pixels using the Douglas-Peucker algorithm [6]. The output is an undirected embedded graph representing the skeleton, including pointers to pixels that comprise each vertex or edge, as well as the density recorded in the density function for each pixel.

4.4 Refinement Steps

4.4.1 Topology Refinement

In the **(3) Topology refinement** step, the undirected skeleton graph is now refined and enhanced by using the continuous trajectory information to infer edge directions and local graph topology. This consists of several sub-steps as follows.

(a) Density-aware map-matching: First, the input trajectories are map-matched onto the skeleton graph using an adaptation of the hidden Markov model-based map-matching algorithm by Thiagarajan et al. [13]. Instead of using uniform transition probabilities, the density information of the grayscale skeleton is used to compute a weight for each edge, that is proportional to the average density of the pixels on that edge. This yields for every input trajectory a path in the skeleton graph that the input trajectory most likely travelled, hence mapping each trajectory to several traversed edges in the graph. It also determines the direction in which the edge is traversed by the trajectory.

(b) Edge pruning: At this point, the graph is considered to be a directed graph. Now, spurious directed edges are pruned by discarding all directed edges with at most one traversed trajectory. For this, only those trajectories are counted for which the portion that has been mapped to the edge is close enough to the edge, according to an RMSD distance that maps sampled trajectory points to the closest point on the edge.

(c) Vertex merging: After the initial graph extraction it may happen that the topology of intersection vertices has not been correctly extracted. A degree-4 vertex, for example, may be represented as two adjacent degree-3 vertices, see Fig. 4.3. The algorithm therefore loops through all short edges, and considers merging the two adjacent vertices by collapsing the edge. The merged vertex location is the average of the original vertex locations. This merge is only performed, if this does not decrease the number of well-matched trajectories. That is, for each trajectory portion that used to be mapped to an edge incident to one of the two original vertices, an RMSD distance is computed that maps sampled trajectory points to the closest point on a new edge incident to the

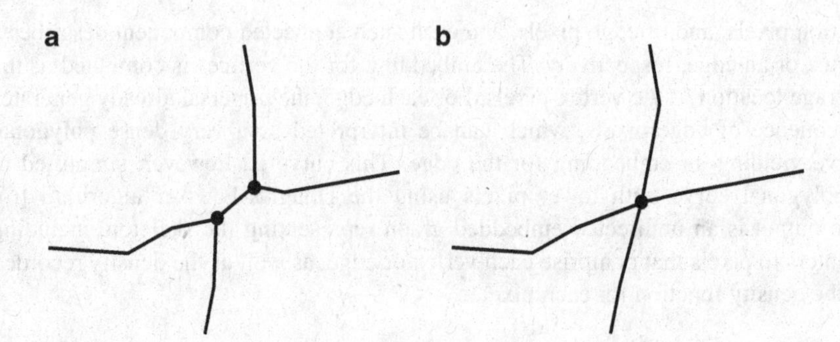

Fig. 4.3 Merging of two intersection vertices. (**a**) Intersection before merging. (**b**) Intersection after merging

merged vertex. And only if this "deformation distance" is small enough the vertex merge is performed.

(d) **Traversal-aware map-matching:** At this point the input trajectories are map-matched once more onto the current pruned graph. The transition probabilities for the hidden Markov model are this time determined for each edge by the number of trajectories that traversed this edge during the first map-matching step (a). This forces trajectories, that used to be mapped to edges that have now been deleted, to be rerouted to other edges.

(e) **Edge pruning:** Another round of edge pruning is performed, using the same algorithm as in step (b). The idea is that the first pruning breaks spurious cycles, which then introduces new spurious spurs, and these spurs will be pruned in the second step.

The directed embedded graph obtained after the topology refinement is then considered as a possible output of this algorithm. It is used in the experiments conducted in the paper [2], and it is the final output of the code provided with the paper.

4.4.2 Geometry Refinement

The final **(4) Geometry refinement** step uses the *k*-means algorithm twice to adjust the embedding of the directed roads, and it adds directed turn lanes and straight lanes for a more detailed representation of intersection vertices. It consists of the following two sub-steps.

(a) **K-means refinement:** The *k*-means algorithm is an iterative point clustering algorithm that repeatedly takes a set of *k* points as the initial set of cluster means, assigns each input point to its closest mean, and then updates the cluster means by calculating the mean of all points assigned to one particular mean. Biagioni and Eriksson apply this idea to adjust the overall geometry, i.e., embedding, of the constructed road network. Each directed edge in the current

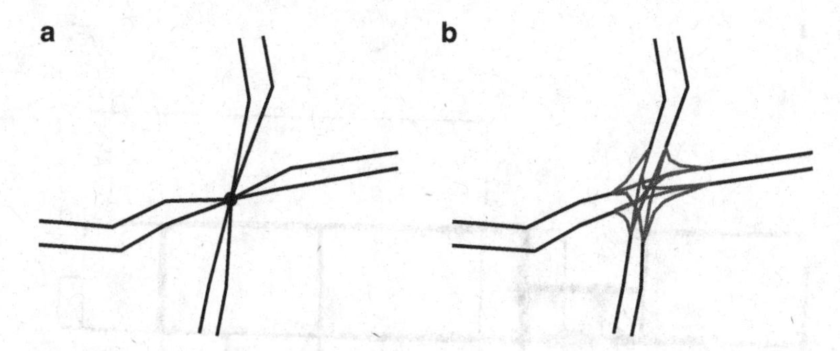

Fig. 4.4 Geometry refinement. (**a**) After *k*-means refinement. (**b**) With added turn lanes and straight lanes

graph, that is embedded as a polygonal curve, is uniformly sampled with points at a certain specified distance. Intersection vertices are represented as single points. Then each measurement from the input trajectories is assigned to the closest sample point, but only those sample points are eligible that lie on the edge that the measurement was matched to during the map-matching in step (d) above. (For this purpose, intersection points are considered to be part of incident edges.) Then the location of each sample point is updated to be the mean of all measurements assigned to it. For bidirected roads this results in a better separation of the two parallel directed edges with one polygonal curve for each direction. The intersection geometry, however, looks somewhat degenerate since it is represented using a single point, see Fig. 4.4a.

(b) Turn lanes: In order to add a more accurate representation of the observed traffic flow at each intersection, all pairwise edge transitions at an intersection are considered that have been witnessed by at least one input trajectory. Then the *k*-means refinement from (a) above is run, using measurements from those trajectory portions that have been map-matched into one of the transition edges. This results in a new representation of an intersection "vertex" as a set of individual turn lanes and straight lanes, which are attached to the graph edges at a constant offset. See Fig. 4.4b for an example of such an intersection refinement.

4.5 Visual Summary

Each step of the pipeline produces a different intermediate map construction result. The implementation that has been provided by the authors does not include the geometry refinement step. An example of the output of the algorithm pipeline is shown in Fig. 4.5. The input data consists of 889 tracks with a total length of 2869 km obtained from university shuttle buses covering an area of $7 \times 4.5\,\text{km}^2$ in

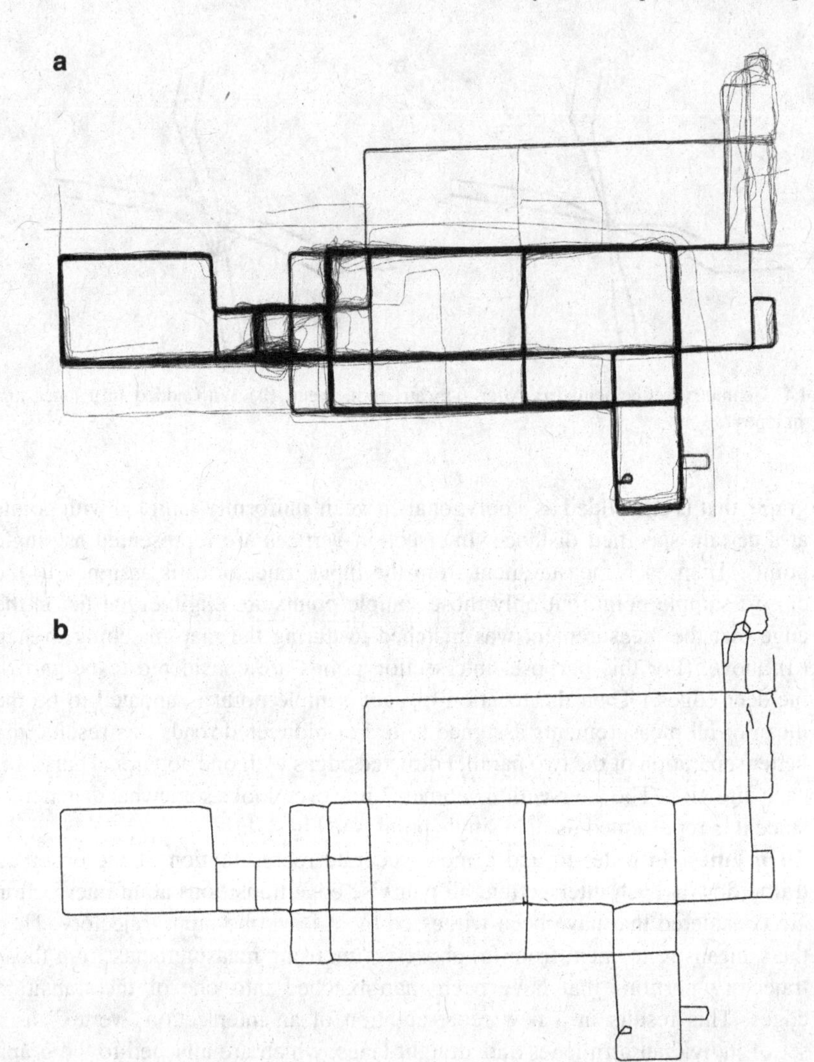

Fig. 4.5 Map construction result after the topology refinement step of Biagioni and Eriksson's [2] pipeline. (**a**) Raw data. (**b**) Constructed map

Chicago [1, 2]. The tracks range from 100 to 363 position samples, with a sampling rate of 1–29 s and an average speed of 33.14 km/h.

References

1. Biagioni, J., Eriksson, J.: Inferring road maps from global positioning system traces: survey and comparative evaluation. Transp. Res. Rec. J. Transp. Res. Board **2291**, 61–71 (2012)
2. Biagioni, J., Eriksson, J.: Map inference in the face of noise and disparity. In: Proceedings of 20th ACM SIGSPATIAL International Conference on Advances in Geographic Information Systems, pp. 79–88 (2012)
3. Chen, C., Cheng, Y.: Roads digital map generation with multi-track GPS data. In: Proceedings of Workshops on Education Technology and Training, and on Geoscience and Remote Sensing, pp. 508–511. IEEE (2008)
4. Davies, J.J., Beresford, A.R., Hopper, A.: Scalable, distributed, real-time map generation. IEEE Pervasive Comput. **5**(4), 47–54 (2006)
5. Dlotko, P., Specogna, R.: Topology preserving thinning of cell complexes. IEEE Trans. Image Process. **23**(10), 4486–4495 (2014)
6. Douglas, D., Peucker, T.: Algorithms for the reduction of the number of points required to represent a digitized line or its caricature. Can. Cartographer **10**(2), 112–122 (1973)
7. Lam, L., Lee, S.W., Suen, C.: Thinning methodologies – a comprehensive survey. IEEE Trans. Pattern Anal. Mach. Intell. **14**(9), 869–885 (1992)
8. Li, Q., Bai, X., Liu, W.: Skeletonization of gray-scale image from incomplete boundaries. In: 15th IEEE International Conference on Image Processing, pp. 877–880 (2008)
9. Qian, K., Cao, S., Bhattacharya, P.: Skeletonization of gray-scale images by gray weighted distance transform. In: Proceedings of SPIE, Visual Information Processing VI, vol. 3074, pp. 224–228 (1997)
10. Saeed, K., Tabędzki, M., Rybnik, M., Adamski, M.: K3m: a universal algorithm for image skeletonization and a review of thinning techniques. Int. J. Appl. Math. Comput. Sci. **20**(2), 317–335 (2010)
11. Shi, W., Shen, S., Liu, Y.: Automatic generation of road network map from massive GPS vehicle trajectories. In: Proceedings of 12th International IEEE Conference on Intelligent Transportation Systems, pp. 48–53 (2009)
12. Steiner, A., Leonhardt, A.: Map generation algorithm using low frequency vehicle position data. In: Proceedings of 90th Annual Meeting of the Transportation Research Board, pp. 1–17 (2011)
13. Thiagarajan, A., Ravindranath, L., LaCurts, K., Madden, S., Balakrishnan, H., Toledo, S., Eriksson, J.: VTrack: accurate, energy-aware road traffic delay estimation using mobile phones. In: Proceedings of 7th ACM Conference on Embedded Networked Sensor Systems, pp. 85–98 (2009)
14. Yim, P.J., Choyke, P.L., Summers, R.M.: Gray-scale skeletonization of small vessels in magnetic resonance angiography. IEEE Trans. Med. Imaging **19**(6), 568–576 (2000)
15. Zhang, T.Y., Suen, C.Y.: A fast parallel algorithm for thinning digital patterns. Commun. ACM **27**(3), 236–239 (1984)

Chapter 5
Datasets

Abstract The best way to get an impression of the capabilities of the various map construction algorithms is to visualize constructed maps side-by-side with the input trajectory data. This chapter showcases map construction results from three different cities and also presents the characteristics of the respective datasets.

5.1 Maps and Trajectories

The performance of map construction algorithms is judged by the quality of the map data they produce. To showcase the performance of the algorithms, trajectory datasets from Chicago (USA), Athens (Greece), and Berlin (Germany) were selected. The respective portions of the maps covered by the trajectory data used in the map construction experiments are shown in Figs. 5.1, 5.2, and 5.3, respectively. The map excerpts are based on OpenStreetMap data [9] and visualized using Toner style maps [11]. Essentially, the ultimate goal for map construction has to be to generate these map datasets based on large amounts of trajectory data. To also quantify this challenge, Table 5.1 gives the number of nodes and edges that comprise those map excerpts. The *Chicago* map covered by the trajectories consists of 11,801 edges and 9429 vertices. It covers an area of $7 \times 4.5\,\text{km}^2$. The edges have a length of 61 km. The underlying *Athens* map portion consists of 3436 edges and 2694 vertices. It covers an area of $2.6 \times 6\,\text{km}^2$. The edges have a length of 193 km. Finally, *Berlin* consists of 6839 edges and 5894 vertices. It covers an area of $6 \times 6\,\text{km}^2$. The edges have a length of 360 km.

The respective trajectory datasets used in this book to showcase and to assess the performance of the various map construction algorithms are shown in Fig. 5.4.

The *Chicago* dataset [2, 3] consists of 889 tracks with a total length of 2869 km (average: 3.22 km and standard deviation: 894 m) obtained from university shuttle buses covering an area of $7 \times 4.5\,\text{km}^2$; the tracks range from 100 to 363 position samples, with a sampling rate of 1–29 s (average: 4 s and standard deviation: 4 s) and an average speed of 33 km/h. The *Athens* dataset consists of 129 tracks with a total length of 443 km (average: 3.82 km and standard deviation: 1.45 km) obtained from school buses covering an area of $2.6 \times 6\,\text{km}^2$; the tracks range from 13 to 47 position samples, with a sampling rate of 20–30 s (average: 34 s and standard deviation: 32 s)

M. Ahmed et al., *Map Construction Algorithms*, DOI 10.1007/978-3-319-25166-0_5

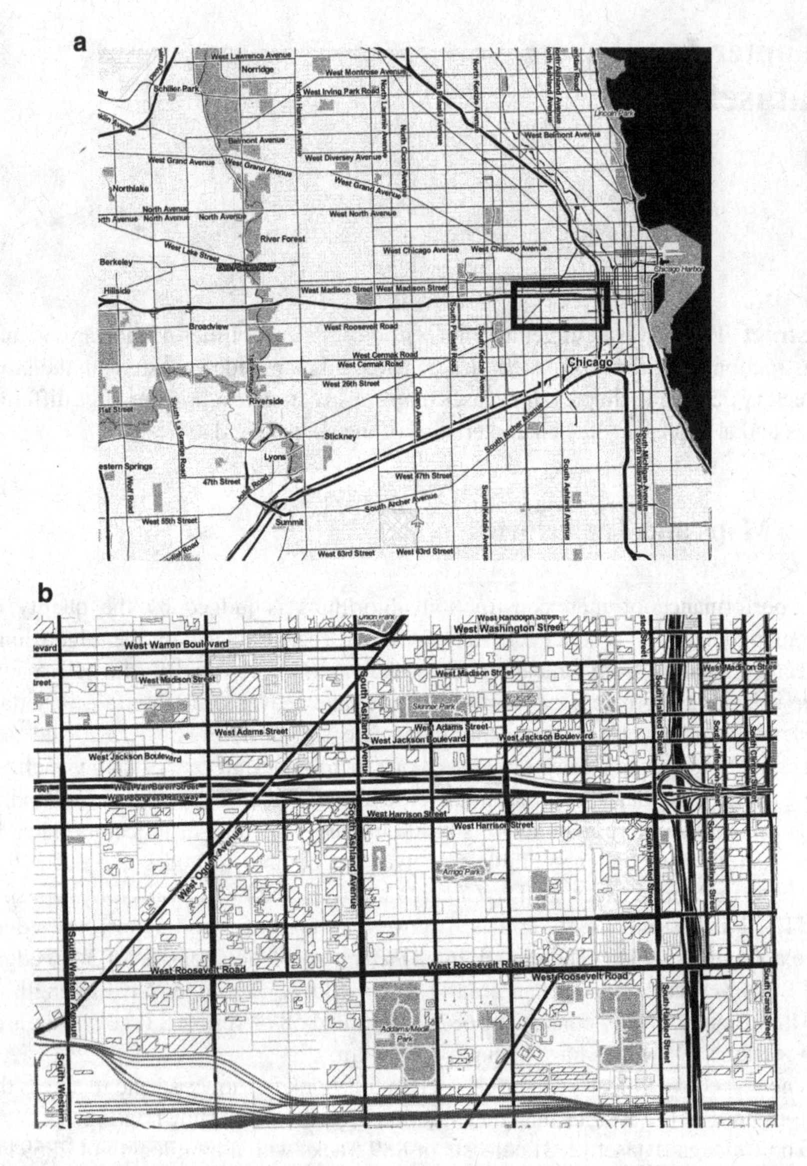

Fig. 5.1 Overview maps and covered areas. (**a**) *Chicago* overview map. (**b**) *Chicago* covered area

and an average speed of 20 km/h. The *Berlin* dataset consists of 26,831 tracks with a total length of 41,116 km (average: 1.53 km and standard deviation: 635 m) obtained from a taxi fleet covering an area of $6 \times 6\,\text{km}^2$; the tracks comprise from 22 up to 58 position samples, with a sampling rate of 15–127 s (average: 42 s and standard deviation: 39 s) and an average speed of 35 km/h. Although covering a similar-sized

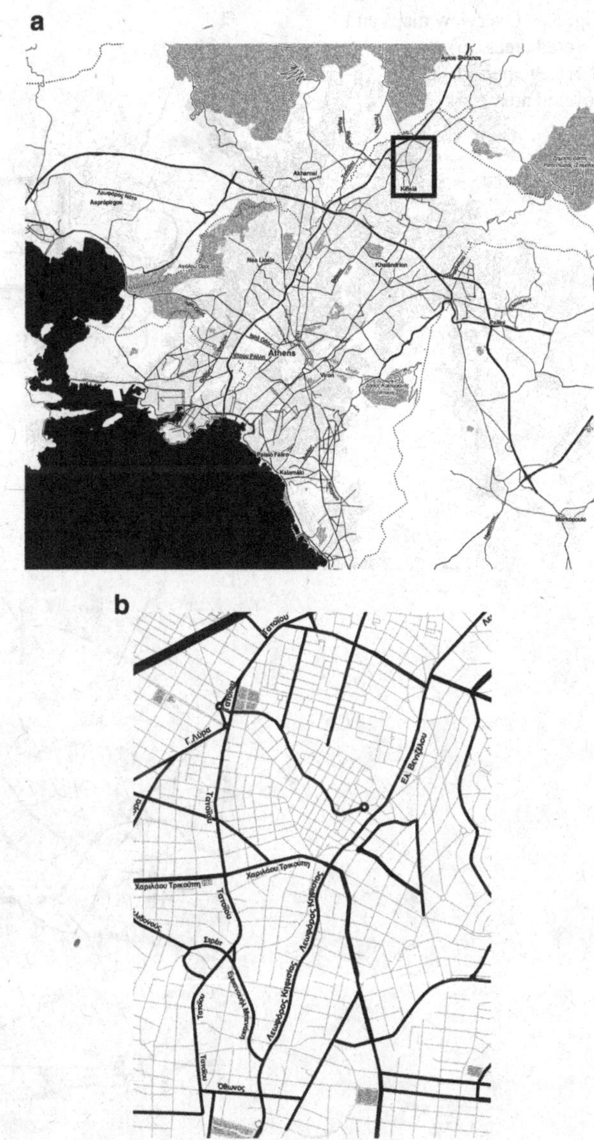

Fig. 5.2 Overview maps and covered areas. (**a**) *Athens* overview map. (**b**) *Athens* covered area

network, the *Berlin* dataset is two orders of magnitude larger than the *Chicago* and *Athens* trajectory data. As such, *Berlin* presents the biggest map construction challenge for our algorithms.

While other publicly available GPS-based vehicle tracking datasets exist, e.g., GeoLife [12] and OpenStreetMap GPX track data [10], the selected range covers the various types of existing datasets produced by different types of vehicles, at varying sampling rates and representing different network sizes.

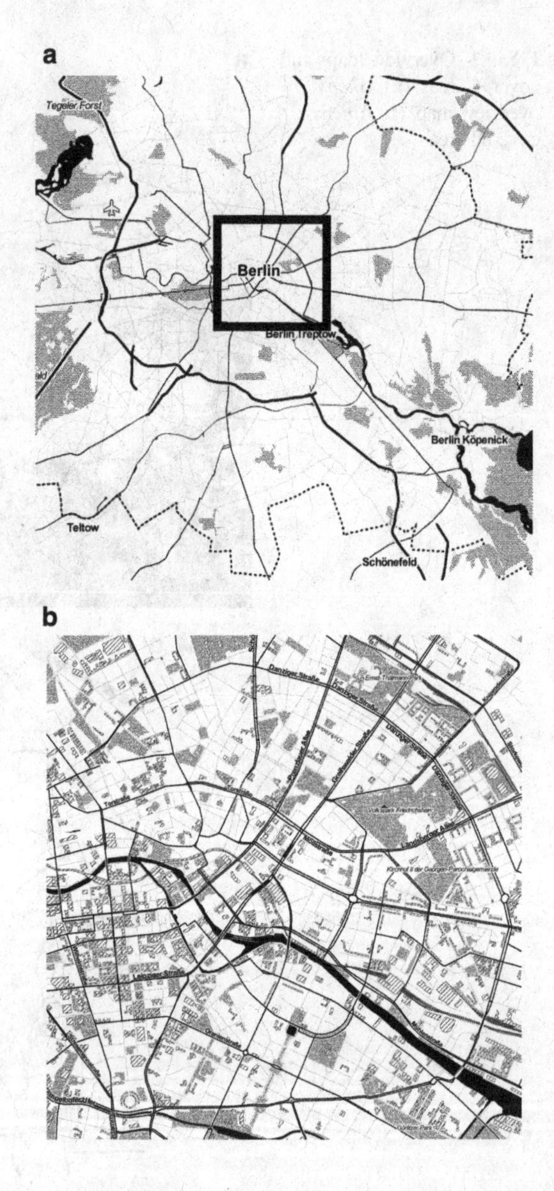

Fig. 5.3 Overview maps and covered areas. (**a**) *Berlin* overview map. (**b**) *Berlin* covered area

5.2 Constructed Maps

Ideally, map construction algorithms use the trajectories of Fig. 5.4 and produce the maps shown in Figs. 5.1, 5.2, and 5.3, respectively.

Table 5.1 Statistics for datasets used

Tracking data	Trajectories	Sampling rate (s)	Trajectory length (km)	Speed (km/h)
Athens	129	34	443	20
Berlin	26,831	42	41,116	35
Chicago	889	4	2,869	33

OSM network	Vertices	Edges	Length (km)	Area (km^2)
Athens	2,694	3,436	193	2.6×6
Berlin	5,894	6,839	360	6×6
Chicago	9,429	11,801	61	7×4.5

The following showcase the map construction results for a range of algorithms discussed in previous chapters. In all visualizations, the OpenStreetMap-derived ground-truth maps are shown in light gray and the generated maps are overlayed in black.

Each of the algorithms uses specific parameter settings. The values of clustering parameters for Ahmed and Wenk's algorithm [1] are 90 m, 170 m and 80 m for *Athens*, *Berlin* and *Chicago*, respectively. For Cao and Krumm's algorithm [4] the clustering parameter is 20 m [4] and the minimum angular difference between two streets at any intersection is 45°. For Edelkamp and Schrödl's algorithm [6] the minimum separation between streets is 50 m and the minimum angular difference between two streets at any intersection is 45°. For the density-based algorithms by Biagioni and Eriksson [3] and by Davies et al. [5] the *minimum density threshold* is set to 50 m and 16 m, respectively. For Karagiorgou and Pfoser's algorithm [8] the values of *direction*, *speed* and *proximity* to extract intersection nodes and to merge trajectories into links are 15°, 40 km/h and 25 m, respectively. All constructed maps were evaluated in Chap. 7 using the distance measures described in Chap. 6.

A summary of the complexities of the constructed maps is shown in Table 5.2. Here, the number of vertices includes vertices of degree two (which may lie on a polygonal curve describing a single edge), the number of edges refers to the number of undirected line segments between these vertices, and the total length refers to the total length of all undirected line segments. It appears that *density-based point clustering algorithms* such as Biagioni and Eriksson [3] and Davies et al. [5] produce maps with lower complexity (fewer number of vertices and edges) but often fail to reconstruct streets that are not traversed frequently enough by the input tracks. In particular, the maps constructed by Davies et al.'s algorithm are very small. On the other hand, the algorithm by Ge et al. [7] subsample all tracks to create a much denser output set, hence the complexity of their constructed maps is always higher.

Map construction algorithms based on *incremental track insertion*, such as Ahmed and Wenk [1] and Cao et al. [4] fail to cluster tracks together when the variability and error associated with the input tracks is large. As a result, the constructed street maps contain multiple edges for a single street, which implies larger values in the total edge length column in Table 5.2.

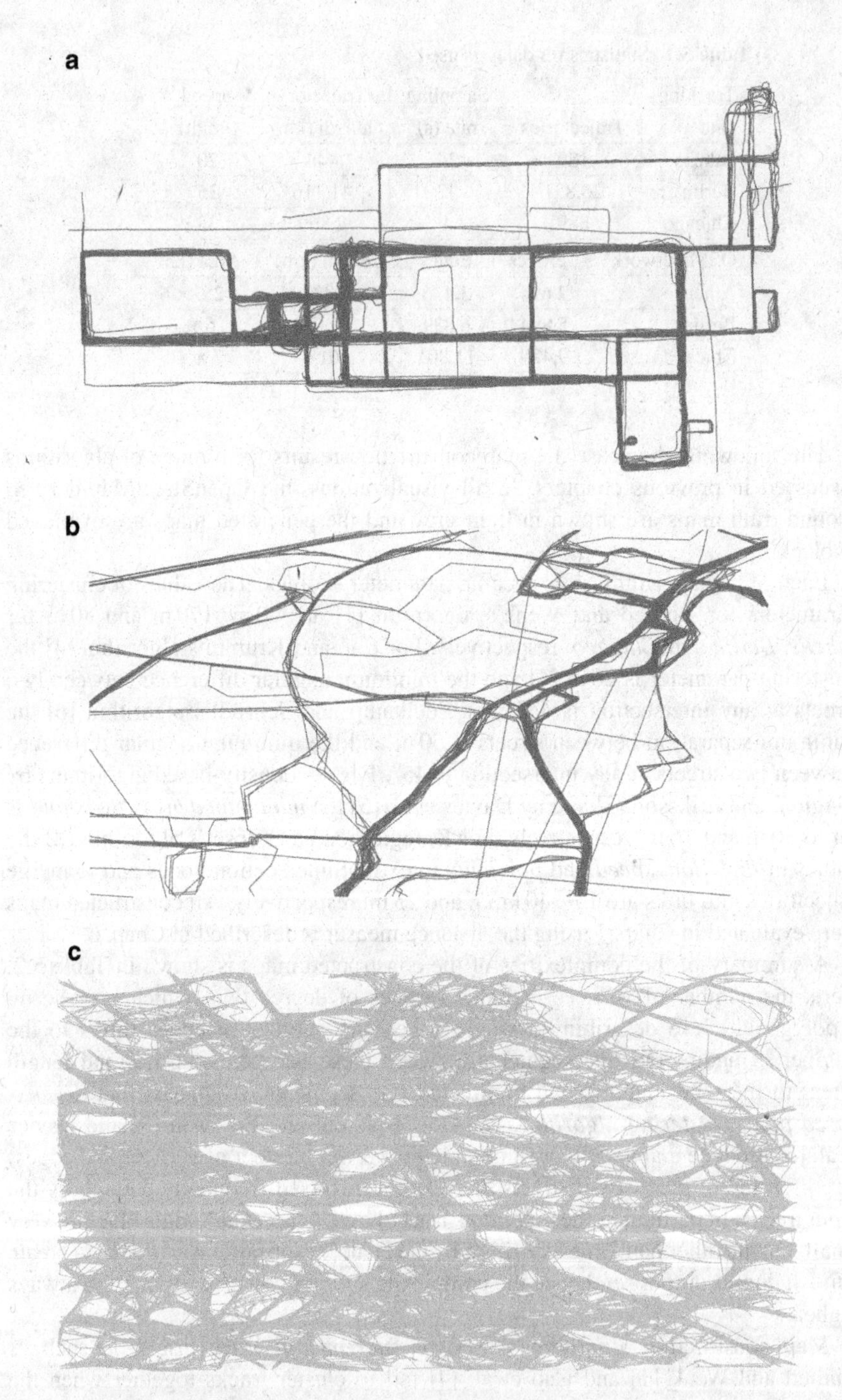

Fig. 5.4 Tracking data. (**a**) *Chicago*. (**b**) *Athens*. (**c**) *Berlin*

Table 5.2 Complexities of the generated maps

Generated map	Vertices	Edges	Length (km)
Athens			
Ahmed	344	378	35
Biagioni	391	398	22
Cao	20	14	3
Davies	209	227	2
Edelkamp	526	1037	197
Ge	1936	1993	23
Karagiorgou	660	637	35
Berlin			
Ahmed	1322	1567	164
Biagioni	5711	5738	184
Ge	15,450	16,136	183
Karagiorgou	2542	2262	161
Chicago			
Ahmed	1195	1286	34
Biagioni	303	322	24
Cao	2092	2948	78
Davies	1277	1310	14
Edelkamp	828	1247	83
Ge	5893	6672	37
Karagiorgou	596	558	26

Several examples of generated maps are shown in Figs. 5.5–5.9 for the datasets of *Chicago*, *Athens* and *Berlin*. Since not all algorithms produced results for all maps, examples of the smaller *Chicago* and *Athens* maps are showcased in Figs. 5.5–5.8. It can be clearly seen that the coverage and quality of the constructed map varies considerably. Four examples for the *Berlin* map are also given in Fig. 5.9. More and continuously updated examples can be found on http://www.mapconstruction.org/.

References

1. Ahmed, M., Wenk, C.: Constructing street networks from GPS trajectories. In: Proceedings of 20th Annual European Symposium on Algorithms, pp. 60–71 (2012)
2. Biagioni, J., Eriksson, J.: Inferring road maps from global positioning system traces: survey and comparative evaluation. Transp. Res. Rec. J. Transp. Res. Board **2291**, 61–71 (2012)
3. Biagioni, J., Eriksson, J.: Map inference in the face of noise and disparity. In: Proceedings of 20th ACM SIGSPATIAL International Conference on Advances in Geographic Information Systems, pp. 79–88 (2012)
4. Cao, L., Krumm, J.: From GPS traces to a routable road map. In: Proceedings of 17th ACM SIGSPATIAL International Conference on Advances in Geographic Information Systems, pp. 3–12 (2009)

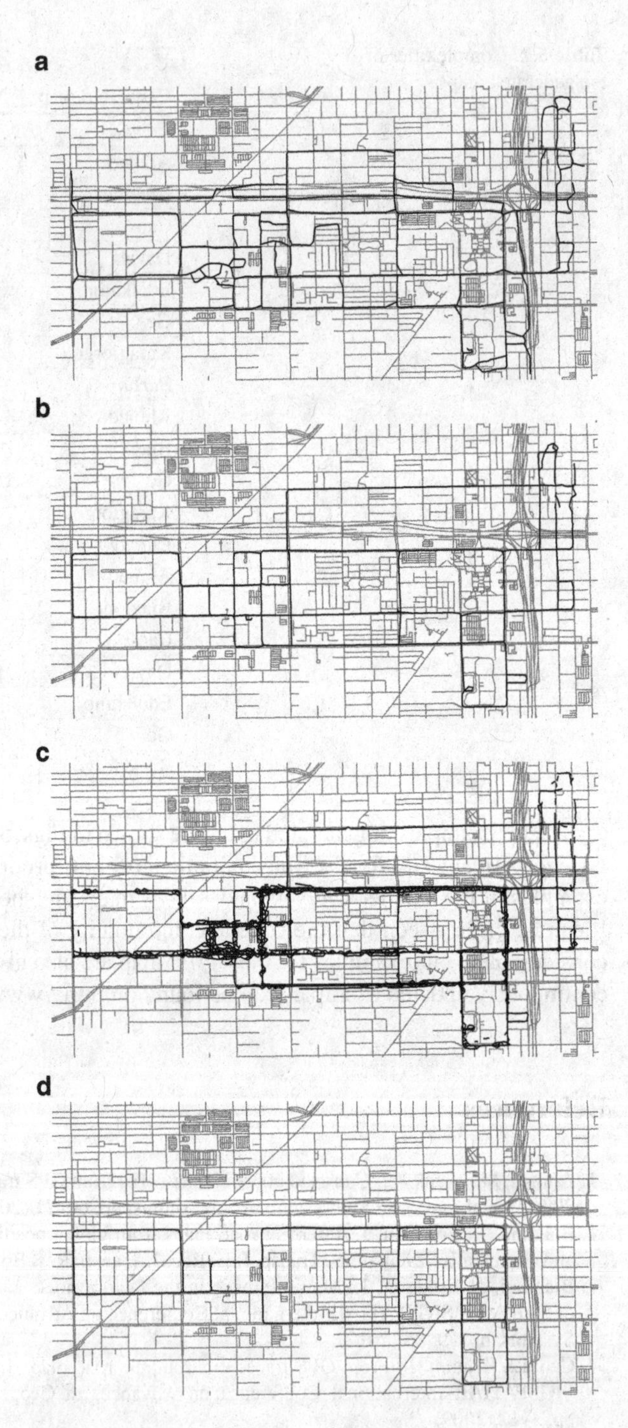

Fig. 5.5 Constructed maps (in *black*) overlayed on ground-truth map (in *gray*) for *Chicago*, part 1. (**a**) Ahmed. (**b**) Biagioni. (**c**) Cao. (**d**) Davies

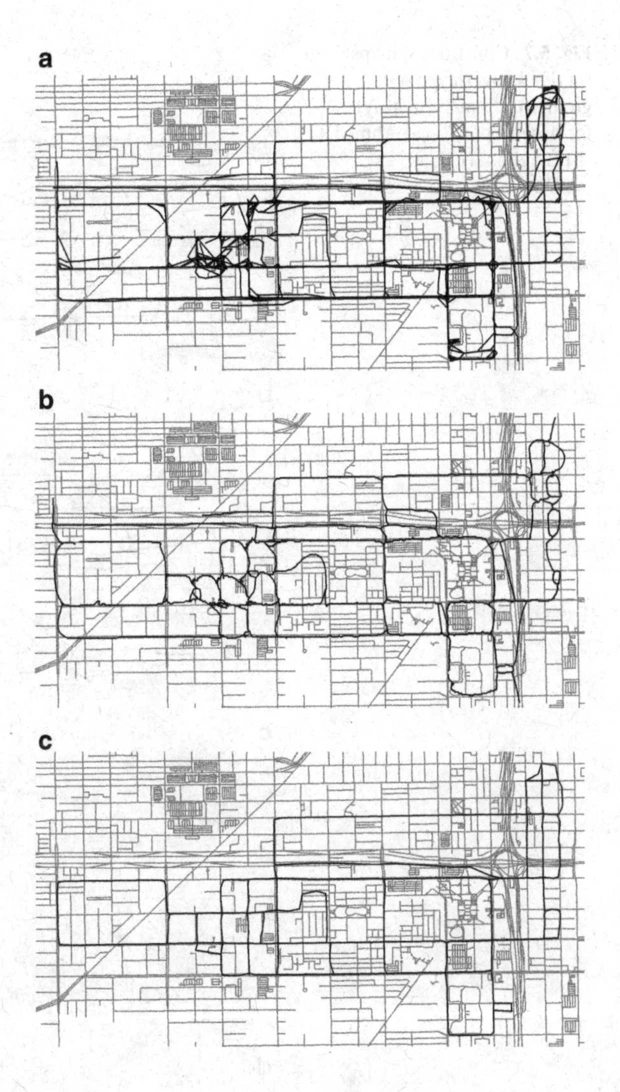

Fig. 5.6 Constructed maps (in *black*) overlayed on ground-truth map (in *gray*) for *Chicago*, part 2. (**a**) Edelkamp. (**b**) Ge. (**c**) Karagiorgou

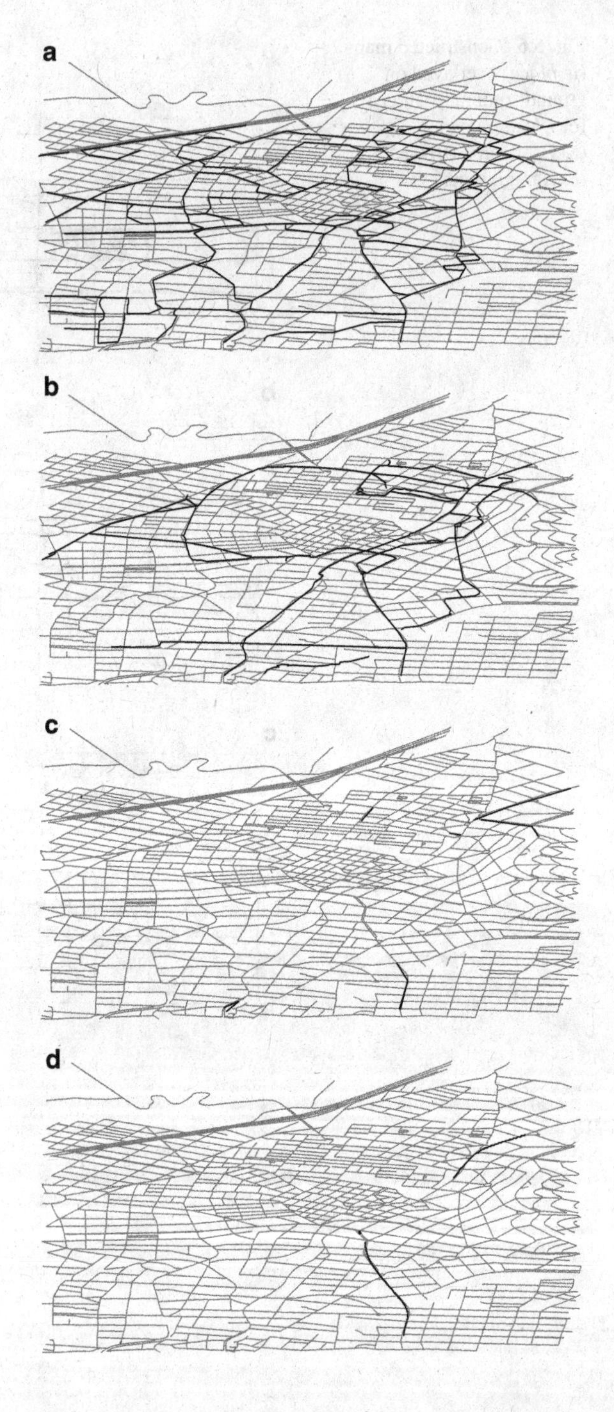

Fig. 5.7 Constructed maps (in *black*) overlayed on ground-truth map (in *gray*) for *Athens*, part 1. (**a**) Ahmed. (**b**) Biagioni. (**c**) Cao. (**d**) Davies

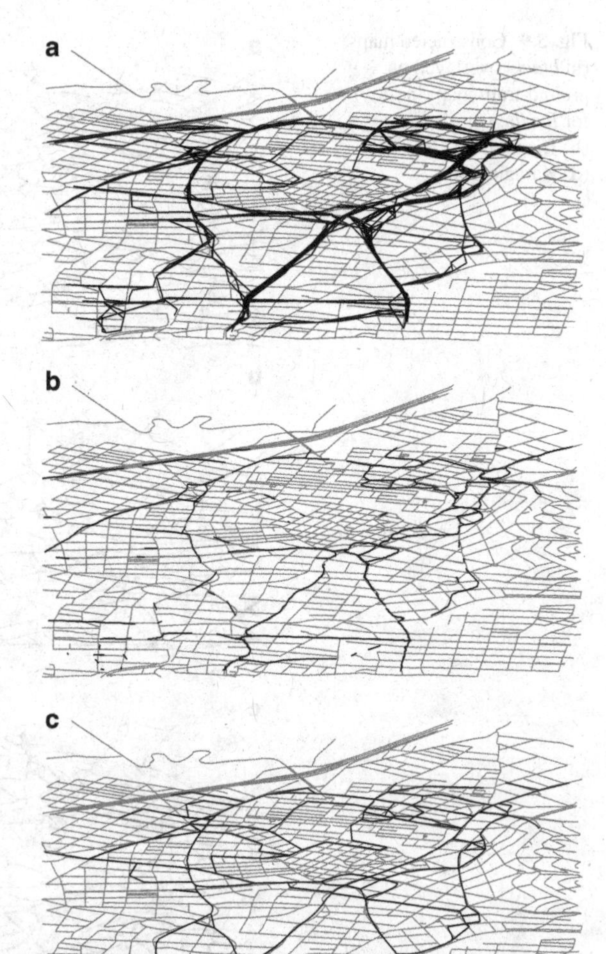

Fig. 5.8 Constructed maps (in *black*) overlayed on ground-truth map (in *gray*) for *Athens*, part 2. (**a**) Edelkamp. (**b**) Ge. (**c**) Karagiorgou

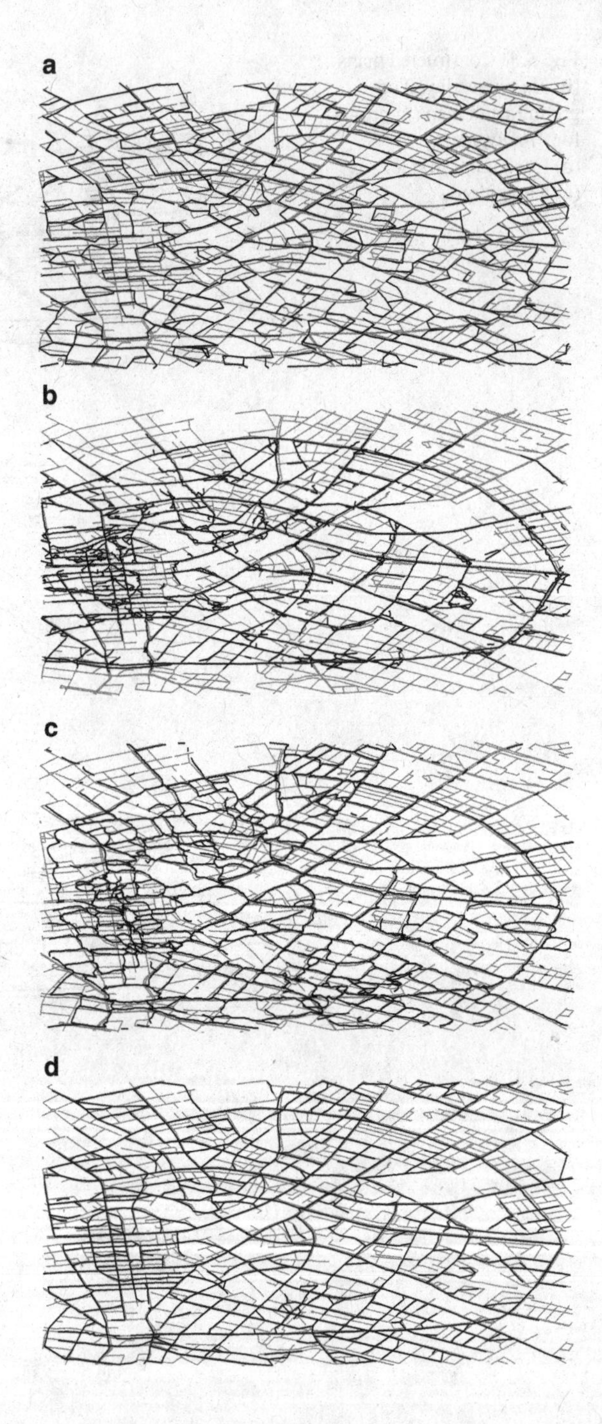

Fig. 5.9 Constructed maps (in *black*) overlayed on ground-truth map (in *gray*) for *Berlin*. (**a**) Ahmed. (**b**) Biagoni. (**c**) Ge. (**d**) Karagiorgou

5. Davies, J.J., Beresford, A.R., Hopper, A.: Scalable, distributed, real-time map generation. IEEE Pervasive Comput. **5**(4), 47–54 (2006)
6. Edelkamp, S., Schrödl, S.: Route planning and map inference with global positioning traces. In: Computer Science in Perspective, pp. 128–151. Springer, Berlin (2003)
7. Ge, X., Safa, I., Belkin, M., Wang, Y.: Data skeletonization via Reeb graphs. In: Proceedings of 25th Annual Conference on Neural Information Processing Systems, pp. 837–845 (2011)
8. Karagiorgou, S., Pfoser, D.: On vehicle tracking data-based road network generation. In: Proceedings of 20th ACM SIGSPATIAL International Conference on Advances in Geographic Information Systems, pp. 89–98 (2012)
9. OpenStreetMap. (2015) http://www.openstreetmap.org/
10. OpenStreetMap Foundation: Bulk GPX track data (2013). http://blog.osmfoundation.org/2013/04/12/bulk-gpx-track-data/
11. Stamen design toner maps. (2015) http://maps.stamen.com/toner/
12. Zheng, Y., Xie, X., Ma, W.Y.: Geolife: a collaborative social networking service among user, location and trajectory. IEEE Data Eng. Bull. **33**(2), 32–39 (2010)

Chapter 6
Quality Measures for Map Comparison

Abstract Map construction algorithms are usually evaluated by comparing the constructed map to a ground-truth map. In this chapter, several quality measures that have been used for map comparison are described. While these quality measures are related to the comparison of abstract graphs, map comparison is a more specialized problem since the common spatial embedding is a key property of road networks that captures the similarity of travel. The distance measures generally work for undirected or directed embedded graph models, and they include partial distances that match the constructed graph to a subset of the ground-truth. The general approaches include point set-based distances, path-based distances, and distances that compare the local topology of the graphs. In addition to distance measures, the concept of local distance signatures is introduced in order to visualize local differences and to locate the cause of large distances.

6.1 Introduction

Map construction algorithms in the literature are usually evaluated by comparing the constructed map to a ground-truth map. While street map data is available from professional vendors or from open sources such as OpenStreetMap, it is tricky, however, to restrict existing ground-truth maps in an unbiased way to cover only the portion that has been traversed by the input set of trajectories. Some quality measures therefore consider a partial matching of the reconstructed map to a larger ground-truth map that is a superset of the desired output.

Quality measures between two street maps are closely related to comparing graphs. While graph comparison algorithms are usually based on structural properties of the graphs, such as their degree distribution, or their local connectivity properties, they ignore any spatial embedding of the graphs. The spatial embedding, however, is a key property of road networks since the similarity of travel over two road networks is intimately tied to the specific spatial embedding.

In this chapter a street map is modeled as an undirected geometric planar graph embedded in the plane. That means, each vertex is embedded as a point in the plane, each edge is a curve in the plane, and no two embedded edges intersect. Some of the presented distance measures can also be applied to directed geometric graphs. In the literature, these are the two most common street map models. Note that planarity of the graph is often assumed, even though this does not allow to model bridges or

M. Ahmed et al., *Map Construction Algorithms*, DOI 10.1007/978-3-319-25166-0_6

overpasses. Additional information such as multiple lanes, turn restrictions, mean speed, etc. are sometimes computed in an additional post-processing step, see e.g. [7, 8, 11, 12, 21], but an undirected (or sometimes directed) embedded graph is usually computed as a first step.

In this chapter, five different quality measures are presented: The directed and undirected Hausdorff distances [4], the path-based distance [3], the shortest path-based distance [17], the graph sampling-based distance [6], and the local homology distance [1]. Each quality measure captures different properties of the graphs. In addition to measuring the quality of a map construction algorithm with a single distance value, the concept of local distance signatures is presented. This allows one to locate the differences in the two graphs, and it can be used to produce heat maps on the graphs to visualize those differences. A comprehensive comparison of map construction algorithms that also includes an experimental quality assessment can be found in [2].

6.2 Ground-Truth Maps

There are two key ingredients for evaluating the quality of a constructed map: (1) the availability of an adequate ground-truth map G as part of the benchmark data and (2) a quality measure used to evaluate the similarity between the constructed map C and the ground-truth map G.

There are essentially two cases of what can be considered as a ground-truth map G. Ideally, G is the underlying map consisting of all streets, and only those streets, that have been traversed by the entities that generated the set of input tracks. If such a G was available, then a suitable quality measure would compare C to all of G and in the ideal case C would equal G. However, in practice, it is hard to obtain an unbiased ground-truth map that exactly corresponds to the coverage of the tracking data. This non-trivial task has been addressed in the past by pruning the ground-truth either manually, by proximity to the tracking data, or by map-matching the tracking data to the map [6, 7, 17, 18]. This results in graph topologies influenced by human judgment, or by cropping behaviors of different pruning algorithms, and hence clearly introduces an undesired bias.

On the other hand, it is much easier to obtain an unbiased ground-truth map that contains a *superset* of all the streets covered by the input tracks. Good examples are street maps available from proprietary vendors, or from OpenStreetMap. Therefore, if G is a superset, some *asymmetric* quality measures attempt to *partially* match C to G. In the most general scenario, however, the constructed map C may even contain additional streets that are not present in G. This may be due to newly constructed roads, additional roads of other road categories (small neighborhood roads for example), or simply due to inconsistencies of the coverage of G and C. In general, it may therefore be desirable to design *symmetric* distance measures that match large portions of C and G but neither one of them entirely.

6.3 Quality Measures

In the graph theory literature, there are various distance measures for comparing two abstract graphs that do not necessarily have a geometric embedding [10, 16, 20]. Most closely related to street map comparison are the subgraph isomorphism problem and the maximum common isomorphic subgraph problem, both of which are NP-complete. These however rely on one-to-one mappings of graphs or subgraphs, and they do not take any geometric embedding into account. Graph edit distance [15, 22] is a way to handle noise by seeking a sequence of edit operations to transform one graph into the other, however it is NP-hard as well. Cheong et al. [9] consider a graph edit distance for geometric graphs (embedded in two different coordinate systems, however), and also show that it is NP-hard to compute.

For comparing street maps, distance measures based on *point sets*, *sets of paths*, or based on *local topology* have been proposed. A street map is modeled as an undirected (planar) graph embedded in the plane. *Point set-based distance measures* treat each graph as the set of points in the plane that is covered by the embedding of its vertices and edges. The idea is then to compute a distance between the two point sets. A straightforward distance measure for point sets are the directed and undirected Hausdorff distances [4]. The main drawback of such an approach is that it does not use the topological structure of the graph. Biagioni and Eriksson [6, 18] use two distance measures that essentially both use a variant of a partial one-to-one matching that is based on sampling both graphs densely. The two distance measures compare the total number of matched sample points to the total number of sample points in the graph, thus providing a measure of how much of the graph has been matched. They do require though to have as input a ground-truth graph that closely resembles the underlying map and not a superset.

For *path-based distance measures* on the other hand, the underlying idea is to represent the graphs by sets of paths, and then to define a distance measure based on distances between the paths. This captures some of the topological information in the graphs and paths are of importance for street maps in particular since the latter are often used for routing applications for which similar connectivity is desirable. Mondzech and Sester [19] use shortest paths to compare the suitability of two road networks for pedestrian navigation by considering basic properties such as respective path length. Karagiorgou and Pfoser [17] also use shortest paths, but with the main goal to assess the similarity of road network graphs. Computing random sets of start and end nodes, the computed paths are compared using discrete Fréchet distance and the average vertical distance. Using those sets of distances, a global network similarity measure is derived. Ahmed et al. [3] cover the networks to be compared with paths of link-length k and map-match the paths to the other graph using the Fréchet distance. In another effort, Ahmed et al. [1] compare the *local topology* of the two graphs. They introduce a local homology distance measure that uses the bottleneck distance between persistence diagrams. Note that Biagioni and Eriksson's sampling-based distance measure [6, 18] also aims at capturing the local topology using discrete point sets.

Both the local homology distance and the path-based distance apply the concept of *local signature* to identify *how* and *where* two graphs differ. Locally, the set of paths through a vertex, or the persistence diagrams of sub-graphs within a small region around a vertex can be compared efficiently, and the results can be visualized as a heat map on one of the graphs to identify and highlight local differences in the graphs.

6.3.1 Directed Hausdorff Distance

The *directed Hausdorff distance* between two compact sets of points A, B is defined as $\overrightarrow{d_H}(A, B) = \max_{a \in A} \min_{b \in B} d(a, b)$. Here, $d(a, b)$ is usually the Euclidean distance between two points a and b. Intuitively, the directed Hausdorff distance assigns to every point $a \in A$ its nearest neighbor $b \in B$ and takes the maximum of all distances between assigned points. In order to compare two graphs, each graph is identified as the set of points that is covered by the embedding of all its vertices and edges. If the directed Hausdorff distance from graph C to graph G is at most ε, this means that for every point on any edge, or vertex of C there is a point on G at distance at most ε. Or equivalently, every point of C is contained in the Minkowski sum of G with a disk of radius ε; the Minkowski sum intuitively "fattens" G by "drawing" each of its edges with a thick circular pen. This distance measure gives a notion about spatial distance for graphs. If C is the constructed graph and G is the ground-truth, the smaller the distance from C to G, the closer the graph C to G. If desired, a symmetric variant of this distance can be computed, the *(undirected) Hausdorff distance* is defined as $d_H(A, B) = \max\{\overrightarrow{d_H}(A, B), \overrightarrow{d_H}(B, A)\}$.

6.3.2 Path-Based Distance

For the path-based distance [3], the embedded graphs are represented as sets of paths. The distance between two such path sets is defined as their directed Hausdorff distance, and for the underlying distance between paths the authors used the Fréchet distance between paths. For curves $f, g : [0, 1] \to \mathbb{R}^d$, the Fréchet distance is defined as

$$\delta_F(f, g) = \inf_{\alpha,\beta:[0,1]\to[0,1]} \max_{t\in[0,1]} d(f(\alpha(t)), g(\beta(t))),$$

where α, β range over continuous, surjective and non decreasing reparametrizations. A common intuition is to explain the Fréchet distance as the minimum leash length required such that a man and dog can continuously walk on the two curves from beginning to end in a monotonic way.

Definition 6.1 (Path-Based Distance). Let C and G be two planar geometric graphs, and let $\pi_C \subseteq \Pi_C$ and $\pi_G \subseteq \Pi_G$, where Π_G is the set of all paths in G. The directed *path-based distance* between these path sets is defined as:

$$\overrightarrow{d}_{path}(\pi_C, \pi_G) = \max_{p_C \in \pi_C} \min_{p_G \in \pi_G} \delta_F(p_C, p_G).$$

The undirected version of the distance $d_{path}(\pi_C, \pi_G)$ is defined to be the maximum of the two directional distances $\overrightarrow{d}_{path}(\pi_C, \pi_G)$ and $\overrightarrow{d}_{path}(\pi_G, \pi_C)$, similar to the undirected Hausdorff distance. Like the Hausdorff distance, the path-based distance is not symmetric, i.e., $\overrightarrow{d}_{path}(\pi_C, \pi_G) \neq \overrightarrow{d}_{path}(\pi_G, \pi_C)$. This anti-symmetry, however, is desirable if G is the ground-truth that covers a superset of what should be covered by constructed map C.

Ideally, π_C and π_G should be the set of all paths in C and G; either starting in vertices or in the interior of an edge. However, such a set would be exponential or even infinite in size. In [3] the authors showed that $\overrightarrow{d}_{path}(\pi_C, \pi_G)$ can be approximated using $\overrightarrow{d}_{path}(\Pi_C^3, \Pi_G)$ in polynomial time using the map-matching algorithm of [5], if C consists of vertices with degree ≥ 4 and the vertices are well-separated. These results are shown by mapping each short path individually from C to G, and then performing surgery to glue the paths together and bound the distortion that arises. See Fig. 6.1 for an example. Here, Π_C is the set of all paths and Π_C^3 is the set of all link-3 paths of C. A link-k path consists of k edges, where vertices of degree two in the graph are not counted as vertices.

The *local signature* of an edge $e \in C$ is defined as $\Delta_e = \overrightarrow{d}_{path}(\Pi_{C_e}, \Pi_G)$, where Π_{C_e} is a set of paths that contains e. Based on the value of these signatures, one can identify which streets of C are very similar to the streets of G and which are not.

The degree assumption on G is only a technical requirement for the theoretical quality guarantees, and the authors have shown [3] that even if the graph does not satisfy the assumptions, it is possible to compare graphs and locate differences between graphs using local signatures computed based on the path based distance

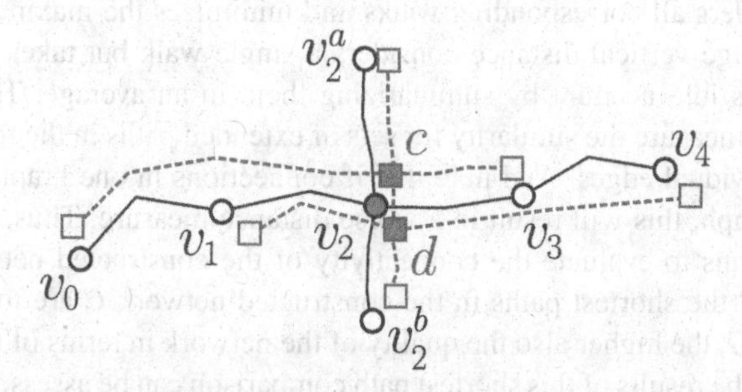

Fig. 6.1 Graph C is shown in *solid lines* and graph G in *dashed lines*. Link-length three paths in C such as v_0, v_1, v_2, v_3 have corresponding paths in G. By performing surgery on these, a path in G can be constructed that corresponds to the link-length four path v_0, v_1, v_2, v_3, v_4 in C

measure. The authors also showed that it suffices to compute $\overrightarrow{d}_{path}(\Pi_C^1, \Pi_H)$ and $\overrightarrow{d}_{path}(\Pi_C^2, \Pi_G)$ in order to identify differences (missing streets/turns), which are most common in street-maps.

Similar to directed Hausdorff distance, the lower the value of $\overrightarrow{d}_{path}(\Pi_C, \Pi_G)$ the more closely the constructed map C resembles the ground-truth map G.

6.3.3 Shortest Path-Based Distance

The quality of constructed maps can be assessed by indirectly comparing them to ground-truth maps using the shortest path-based distance as proposed by Karagiorgou et al. [17]. A random set of origin-destination pairs is used to compute a respective set of shortest paths in both the constructed and the ground-truth map. Assuming a perfect map construction result, the shortest paths of the constructed map should match the path in the ground-truth map. Since constructed maps typically do not match the ground-truth map, the degree of dissimilarity of the shortest paths is an indicator of the quality of the constructed map.

Before being able to compute shortest paths, one needs to establish the ground-truth map, i.e., a subset of the road network that corresponds to the coverage of the trajectory data used to construct a map. The ground-truth map is found by initially creating buffer regions around all edges of the complete road network. These "fattened" edges are then intersected with the trajectory data. The ground-truth map consists then of all edges whose buffer regions contain at least one trajectory segment.

Given the constructed map C and the ground-truth map G, a random set of origin-destination pairs is chosen and respective shortest paths are computed in the C and G, respectively. The geometric difference/similarity between the sets of shortest paths is used to assess the similarity between C and G, and thus the quality of the constructed network. To compare the shortest paths the discrete Fréchet distance and the average vertical distance measures are used. The discrete Fréchet distance considers all corresponding walks and minimizes the maximum distance, while the average vertical distance considers a single walk but takes all distances along the paths into account by summarizing them in an average. This approach enables one to measure the similarity for sets of extended paths in the road networks instead of individual edges. And note that if connections in one graph are missing in the other graph, this will result in a worse distance measure. Thus, this distance provides a means to evaluate the connectivity of the constructed network C. The more "similar" the shortest paths in the constructed network C are to the ground-truth network G, the higher also the quality of the network in terms of topology and connectivity. The results of this shortest path comparison can be assessed by plotting the distance of all paths against each other and can be summarized by computing average distances for a set of paths for the two networks that are compared.

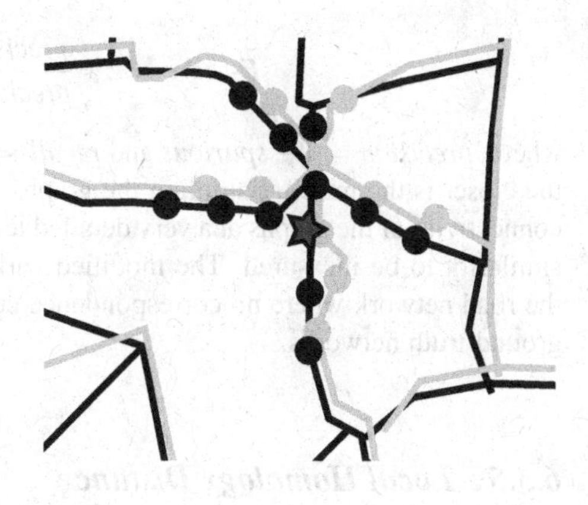

Fig. 6.2 For the graph sampling-based distance the sampled points on the *dark* graph need to be matched to sampled points on the *light* graph using a one-to-one bottleneck matching. The seed location is marked with a *star*

6.3.4 Graph Sampling-Based Distance

Biagioni and Eriksson [6] introduce a graph sampling-based distance measure in order to evaluate the geometry and the topology of the constructed maps represented by graphs. The main idea is as follows. Starting from a random street location, explore the topology of the graphs by placing point samples on each graph during a graph traversal outward within a maximum radius. This produces two sets of locations, which are essentially spatial samples of a local graph neighborhood. These two point sets are compared using one-to-one bottleneck matching [14] and counting the unmatched points in each set. See Fig. 6.2 for an illustration. Note that the graph traversal can take directions of edges into account if desired; in [6] the authors apply this distance measure to directed graphs, while in [7] the authors apply it to undirected graphs.

The sampling process is repeated for several random seed locations. For the bottleneck matching, the sample points on one graph can be considered as "marbles" and on the other graph as "holes". The algorithm considers one-to-one matchings between the point-sets and only allows points to be matched that are at a distance less than a given threshold. Intuitively, if a marble lands close to a hole it falls in, marbles that are too far from a hole remain where they land, and holes with no marbles nearby remain empty. If one of the graphs is the ground-truth, this difference represents the accuracy of the other graph. Counting the number of unmatched marbles and empty holes quantifies the accuracy of the generated road network with respect to the ground-truth according to two scores. The first score is the proportion of spurious marbles, $spurious = spurious\_marbles / (spurious\_marbles + matched\_marbles)$ and the second score is the proportion of missing locations or empty holes, where $missing = empty\_holes / (empty\_holes + matched\_holes)$.

To produce a combined performance measure from these two values, the well-known F-score is used, which is computed as follows:

$$F\text{-}score = 2 * \frac{precision * recall}{precision + recall}$$

where, $precision = 1 - spurious$ and $recall = 1 - missing$. The higher the F-score, the closer is the match. Sampling the graphs locally is important as it captures the connectivity of the graphs at a very detailed level and thus allows for the topological similarity to be measured. The modified variant presented in [7] ignores parts of the road network where no correspondence could be found between generated and ground-truth networks.

6.3.5 Local Homology Distance

Ahmed et al. [1] present a distance measure that is based on comparing the local persistent homology of two graphs.

Let G be a graph embedded in the plane. While the homology of G describes the connected components as well as the cycles and branching structures present in the graph, it cannot distinguish between different sizes of components and cycles. To capture the topology of G at different scales, the graph is continuously thickened until all cycles are filled in. This results in a *filtration*, i.e., a sequence of growing topological spaces. One can think of this thickening process as parameterized by time. During this process new cycles can appear and existing cycles can disappear. A thickened graph at time t can be described by the sublevel set $[G_i]^t := d_{G_i}^{-1}((-\infty, t])$ of the Euclidean distance function $d_G \colon \mathbb{R}^2 \to \mathbb{R}$ in the plane. Here, $d_G(p)$ is defined as the distance from p to the point closest to it in G. The time at which a cycle appears is called the *birth* of a cycle, and the time at which a cycle is filled in is called the *death* of a cycle. The persistent homology then defines a set of birth-death pairs, which are obtained from homology generators over the whole thickening process (filtration). These birth-death pairs are plotted in a *persistence diagram*. Each such pair is referred to as a *feature*, and the difference between birth and death is called its *persistence*. Cycles with high persistence can be interpreted as important homological features of the road network.

For two given graphs G_1, G_2 embedded in the plane, let $\mathscr{P}(G_1)$ and $\mathscr{P}(G_2)$ be their persistence diagrams, which each are a set of points in the plane. The goal is now to find a correspondence between the homological features in the two diagrams. The idea is to find a one-to-one correspondence between the points in the diagrams, however for technical reasons a matching to the diagonal with equal birth and death time is allowed (these correspond to cycles that appear and then immediately disappear). A popular distance measure used for comparing persistence diagrams is the *bottleneck distance* $W_\infty(\mathscr{P}(G_1), \mathscr{P}(G_2)) := \inf_f ||p - f(p)||_\infty$, where $f \colon \mathscr{P}_1 \to \mathscr{P}_2$ ranges over all bijections; see [13, Chap. VII]. See Fig. 6.3 for an example of two graphs, their persistence diagrams, and the pairing of the persistence points with respect to the bottleneck distance.

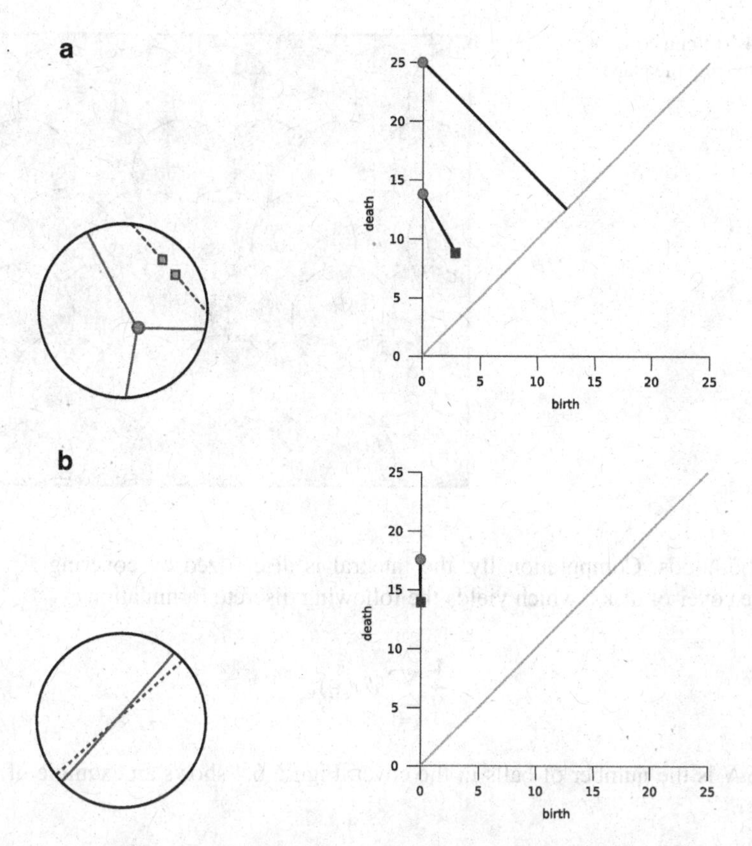

Fig. 6.3 Two examples of local homology distance. The local portions of two graphs (*solid* and *dashed*) on the *left*, and their persistence diagrams on the *right*. Persistence points are paired with *solid lines*. (**a**) Large local homology distance. (**b**) Small local homology distance

In order to capture local features, this distance between persistence diagrams is applied to local portions of the graph. For a given radius $r > 0$ and a point x in the plane, let $\mathscr{P}_{1,r}(x)$ be the persistence diagram of the portion of G_1 within the ball of radius r centered at x. $\mathscr{P}_{2,r}(x)$ is defined analogously for G_2. Then the *local distance signature* $\psi_r: \mathbb{R}^2 \to \mathbb{R}$ is defined by $\psi_r(x) = W_\infty(\mathscr{P}_{1,r}(x), \mathscr{P}_{2,r}(x))$. This signature is then used to define the *local homology distance* as

$$d_r^{LH}(G_1, G_2) = \int_{\mathbb{R}^2} w(x)\psi_r(x)\, dx,$$

where $w: \mathbb{R}^2 \to \mathbb{R}$ is a non-negative weight function that integrates to unity. The local homology distance thus captures local topological similarity between the graphs and integrates over the whole plane in order to capture all local

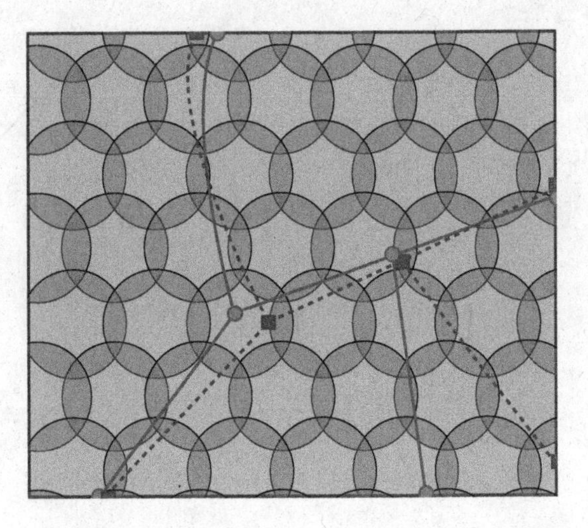

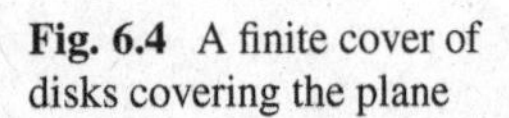

Fig. 6.4 A finite cover of disks covering the plane

neighborhoods. Computationally, the integral is discretized by covering $\mathbb{R}^2$ with a finite cover of disks, which yields the following discrete formulation

$$\frac{1}{N}\sum_{i=1}^{N}\psi_r(x_i),$$

where N is the number of balls in the cover. Figure 6.4 shows an example of such a cover.

6.4 Local Distance Signatures

A distance measure reduces the differences between two maps to a single value, which is useful when assessing and comparing the quality of multiple constructed maps. A related objective to measuring the quality of the constructed map is to *identify* where the constructed map C differs from the ground-truth map G. This requires a means to localize differences and visualize them. One approach is to define a *local distance signature*. For a given point p (or vertex, or edge) on one graph, p defines local neighborhoods in the graph and the other graph, and the signature of p is the distance between the two local graph portions. Since the local portions are defined via points on the graph, a heat map can be plotted on the graph in order to visualize the local distance signature at different points on the graph such as shown in the example of Fig. 6.5.

The concept of local distance signature has been employed for two distance measures, the path-based distance [3] (cf. Sect. 6.3.2) and the local homology distance [1] (cf. Sect. 6.3.5). For the path-based distance, the point p defines the set of short paths through p, and for the local homology distance the point p defines a

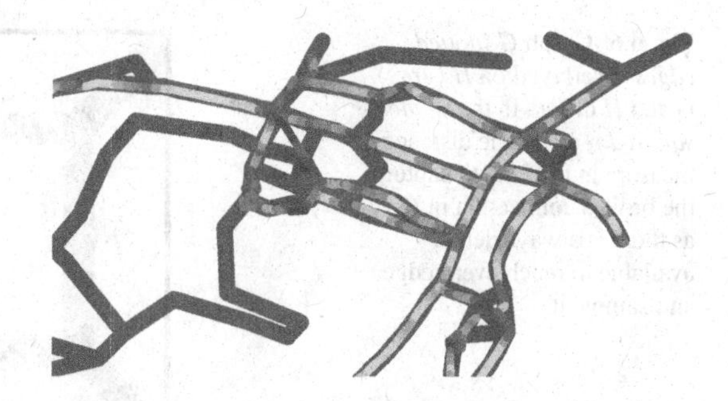

Fig. 6.5 The color-coded street map is overlayed on the *gray* street map. *Colors* show the heat map of the local distance signature of the local homology distance

graph neighborhood, which is used to compute and compare persistence diagrams. Figure 6.5 shows an example for the local homology distance. The algorithm to compute the graph sampling-based distance [7] can also be naturally adapted to compute local distance signatures, if initial seeds are taken to be points on one of the graphs.

6.5 Comparison of Distance Measures

All the distance measures described in Sect. 6.3 capture different properties of graphs. Based on the desired type of similarity, different distance measures can be employed. For example, if one is interested in ensuring that shortest paths in the two graphs are similar, requiring that independent queries produce similar routes, then the shortest path-based distance would be a good choice [17, 19]. If the basic spatial displacement between the two graphs is of importance, without necessarily considering any kind of topology or path similarity, then the directed Hausdorff distance [4] would be appropriate.

On the other hand, the graph sampling-based distance [7] and the local homology distance [1] maximize the use of local topology in comparing graphs. Note that the path-based distance combines topology, connectivity, and spatial similarity by considering all paths. The graph sampling-based distance [7] may, due to the employed sampling strategy, fail to identify local differences. Figure 6.6 shows an example in which the dotted graph has a broken connection in the gray square region, but the graph topology allows all edges to be traversed and sampled nevertheless, and hence the resulting sampled point sets are similar. The path-based distance [3] on the other hand exploits every adjacency transition around a vertex and therefore verifies connectivity. The local homology distance [1] considers continuous cycles and would therefore also detect the different topologies of the two graphs.

Among the presented distance measures, only the graph sampling-based distance [7] ensures one-to-one correspondence between portions (point samples) of the graphs. Therefore, missing streets or extra edges are reflected in the overall score as well as in the local signatures.

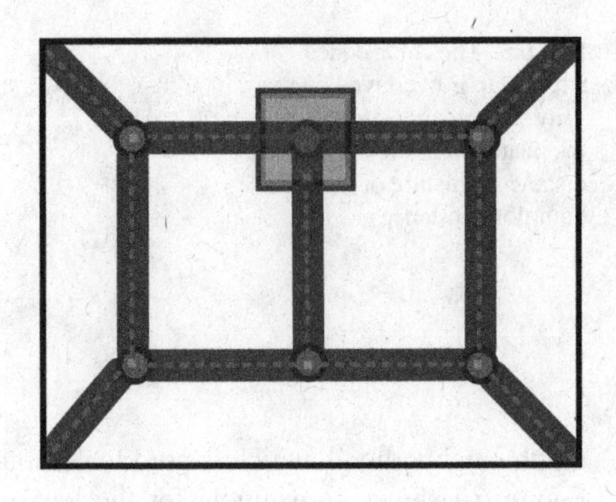

Fig. 6.6 Graph G (*dotted edges*) overlayed on H (*gray*). G and H differs in the *shaded squared region*. The distance measure in [7] fails to capture the broken connection in G, as there is always detour available to reach every edge and sample it

References

1. Ahmed, M., Fasy, B.T., Wenk, C.: Local persistent homology based distance between maps. In: Proceedings of 22nd ACM SIGSPATIAL International Conference on Advances in Geographic Information Systems, pp. 43–52. ACM (2014)
2. Ahmed, M., Karagiorgou, S., Pfoser, D., Wenk, C.: A comparison and evaluation of map construction algorithms using vehicle tracking data. GeoInformatica **19**(3), 601–632 (2015)
3. Ahmed, M., Fasy, B.T., Hickmann, K.S., Wenk, C.: Path-based distance for street map comparison. ACM Trans. Spatial Algorithms Syst. **1**, 28 pp. (2015)
4. Alt, H., Guibas, L.J.: Discrete geometric shapes: matching, interpolation, and approximation – a survey. In: Sack, J.R., Urrutia, J. (eds.) Handbook of Computational Geometry, pp. 121–154. Elsevier, Amsterdam (1999)
5. Alt, H., Efrat, A., Rote, G., Wenk, C.: Matching planar maps. J. Algorithms **49**, 262–283 (2003)
6. Biagioni, J., Eriksson, J.: Inferring road maps from global positioning system traces: survey and comparative evaluation. Transp. Res. Rec. J. Transp. Res. Board **2291**, 61–71 (2012)
7. Biagioni, J., Eriksson, J.: Map inference in the face of noise and disparity. In: Proceedings of 20th ACM SIGSPATIAL International Conference on Advances in Geographic Information Systems, pp. 79–88 (2012)
8. Cao, L., Krumm, J.: From GPS traces to a routable road map. In: Proceedings of 17th ACM SIGSPATIAL International Conference on Advances in Geographic Information Systems, pp. 3–12 (2009)
9. Cheong, O., Gudmundsson, J., Kim, H.S., Schymura, D., Stehn, F.: Measuring the similarity of geometric graphs. In: Proceedings of International Symposium on Experimental Algorithms, pp. 101–112 (2009)
10. Conte, D., Foggia, P., Sansone, C., Vento, M.: Thirty years of graph matching in pattern recognition. Int. J. Pattern Recogn. Artif. Intell. **18**(3), 265–298 (2004)
11. Davies, J.J., Beresford, A.R., Hopper, A.: Scalable, distributed, real-time map generation. IEEE Pervasive Comput. **5**(4), 47–54 (2006)
12. Edelkamp, S., Schrödl, S.: Route planning and map inference with global positioning traces. In: Computer Science in Perspective, pp. 128–151. Springer, Berlin (2003)
13. Edelsbrunner, H., Harer, J.: Computational Topology: An Introduction. AMS, Providence, RI (2010)
14. Efrat, A., Itai, A., Katz, M.J.: Geometry helps in bottleneck matching and related problems. Algorithmica **31**, 1–28 (2001)
15. Gao, X., Xiao, B., Tao, D., Li, X.: A survey of graph edit distance. Pattern Anal. Appl. **13**, 113–129 (2010)

16. Gati, G.: Further annotated bibliography on the isomorphism disease. J. Graph Theory **3**(2), 95–109 (1979)
17. Karagiorgou, S., Pfoser, D.: On vehicle tracking data-based road network generation. In: Proceedings of 20th ACM SIGSPATIAL International Conference on Advances in Geographic Information Systems, pp. 89–98 (2012)
18. Liu, X., Biagioni, J., Eriksson, J., Wang, Y., Forman, G., Zhu, Y.: Mining large-scale, sparse GPS traces for map inference: comparison of approaches. In: Proceedings of 18th ACM SIGKDD International Conference on Knowledge Discovery and Data Mining, pp. 669–677 (2012)
19. Mondzech, J., Sester, M.: Quality analysis of openstreetmap data based on application needs. Cartographica **46**, 115–125 (2011)
20. Read, R.C., Corneil, D.G.: The graph isomorphism disease. J. Graph Theory **1**(4), 339–363 (1977)
21. Schroedl, S., Wagstaff, K., Rogers, S., Langley, P., Wilson, C.: Mining GPS traces for map refinement. Data Min. Knowl. Disc. **9**, 59–87 (2004)
22. Zeng, Z., Tung, A.K.H., Wang, J., Feng, J., Zhou, L.: Comparing stars: on approximating graph edit distance. In: Proceedings of 35th VLDB Conference, pp. 25–36 (2009)

Chapter 7
Evaluation

Abstract Although a visual inspection allows for a simple intuitive assessment of map construction results, providing a quantifiable assessment of the quality has been a considerable challenge. This chapter summarizes the results of a study that compares three map construction algorithms for three different datasets and using four quality measures. The results illustrate the strengths and limitations of the algorithms representing three distinct categories of map construction approaches.

7.1 Introduction

This chapter showcases map construction experiments that were conducted for a range of algorithms, datasets, and evaluation measures described in previous chapters, with the aim to assess the quality of the constructed maps. Three map construction algorithms, each representing a distinct category of map construction approaches, were used in the experiments: The density-based point clustering algorithm by Biagioni and Eriksson [6], the incremental track insertion algorithm by Ahmed and Wenk [1], and the intersection linking algorithm by Karagiorgou and Pfoser [7]. The implementations use C, Java, Python, and Matlab, and are available on http://www.mapconstruction.org/. For evaluation purposes, only the underlying undirected graph representations that were computed by the algorithms were used; any additional information such as directions or other annotations were dropped. The data sets used for the experiments consist of trajectory data from *Chicago*, *Athens*, and *Berlin*, as well as map excerpts from OpenStreetMap data [8], as described in Chap. 5. The quality measures used for evaluation are the directed Hausdorff distance [4], the path-based distance [3], the shortest path-based distance [7], the graph sampling-based distance [5], and the local homology distance [2]. The experiments have been performed by the authors and the implementations have been made available on http://www.mapconstruction.org/. Given (1) the differences in code base, (2) the goal to construct small-scale maps from GPS trajectories, and (3) the fact that all implementations are academic prototypes, the characteristics of the algorithms were not assessed by means of a performance study or theoretical analysis. The running times of the algorithms range from 10 min to 20 h for smaller data sets such as for the *Chicago* dataset. And for the larger data sets such as the *Berlin* dataset, the running times range from 2 h to 4 d.

M. Ahmed et al., *Map Construction Algorithms*, DOI 10.1007/978-3-319-25166-0_7

7.2 Path-Based Distance and Directed Hausdorff Distance

For the path-based distance all paths of link-length 3 were generated for each constructed map. For each path, a map-matching was performed to compute the Fréchet distance between the path and the ground-truth map. The minimum, maximum, median, and average of all the obtained distances were computed. The $d\%$-distances, defined as the maximum of the distances after removing the $d\%$ largest distances ("outliers"), were also computed. The *directed Hausdorff distance* was computed from the constructed maps to the ground-truth maps. Here each map is considered as the set of all points covered by the vertices and edges of its embedding. The results are summarized in Table 7.1. The maps constructed using the algorithms by Karagiorgou and Pfoser [7] and by Biagioni and Eriksson [6] generally have the best path-based distance.

For further analysis of the results, the *Chicago* dataset was selected as it was one of the first datasets used in respective comparison studies. From Table 7.1 one can see that the path-based distance and the directed Hausdorff distance are smaller for the generated maps by Biagioni and Eriksson and by Karagiorgou and Pfoser compared to maps generated using other algorithms. Although the algorithm of Ahmed and Wenk produces maps with good coverage, their path-based distances are larger. This may be due to the fact that the implementation does not include the last step of their algorithm which would compute a minimum-link representative edge. Such a more aggressive averaging technique should help cope with noise in the input tracks.

To illustrate the appropriateness of the path-based distance, consider the path in Fig. 7.1 from the map generated by Biagioni and Eriksson. This is an example where the Fréchet distance is more effective than any point-based distance. As Fréchet distance ensures continuous mapping, the whole path needs to be matched with the bottom horizontal edge of the ground-truth map. The Fréchet distance for this path is 71 m. For the same path, the Hausdorff distance is 53 m, as this only requires for each point on the path to have a point on the graph close-by. Thus, in order to evaluate the connectivity of a map, the Fréchet distance is more suitable than any point-based measure.

In addition, if desired one can discard outliers by computing the $d\%$-distance. Figure 7.2 shows the distribution of both the path-based distance and the directed Hausdorff distance for the map constructed by Biagioni and Eriksson. In both cases, a very small number of paths have the maximum distance, and the distances for most of the paths are distributed within a small range. Removing only 5% of the outliers (largest distances) brings the path-based distance from 71 m (max) to 38 m and the directed Hausdorff distance from 53 m (max) to 25 m.

Figure 7.3 uses the local distance signature of the path-based distance to visualize the distances in the maps. For each edge, a set of all link-length 3 paths containing this edge are considered, and the path-based distance of this path-set is computed

Table 7.1 Path-based distance and directed Hausdorff distance evaluation

Generated map	Path-based distance (m)								Directed Hausdorff distance (m)							
	min	max	med	avg	2 %	5 %	10 %	15 %	min	max	med	avg	2 %	5 %	10 %	15 %
Athens																
Ahmed	9	224	45	52	101	101	81	72	1	82	25	26	82	54	46	40
Biagioni	5	73	35	36	67	66	61	57	3	74	19	20	47	43	31	31
Karagiorgou	7	229	32	38	113	68	59	57	2	84	14	17	54	40	33	30
Berlin																
Ahmed	9	540	66	74	207	147	120	107	1	219	30	33	95	70	60	53
Biagioni	2	228	43	48	111	93	81	75	1	452	13	24	132	95	55	41
Karagiorgou	4	306	28	37	120	85	65	52	1	232	14	18	59	42	34	30
Chicago																
Ahmed	7	201	35	42	127	100	85	76	1	81	14	19	72	59	43	35
Biagioni	3	71	15	18	71	38	27	26	2	53	9	11	29	25	23	17
Karagiorgou	3	89	15	23	72	72	65	51	1	48	7	8	41	23	15	13

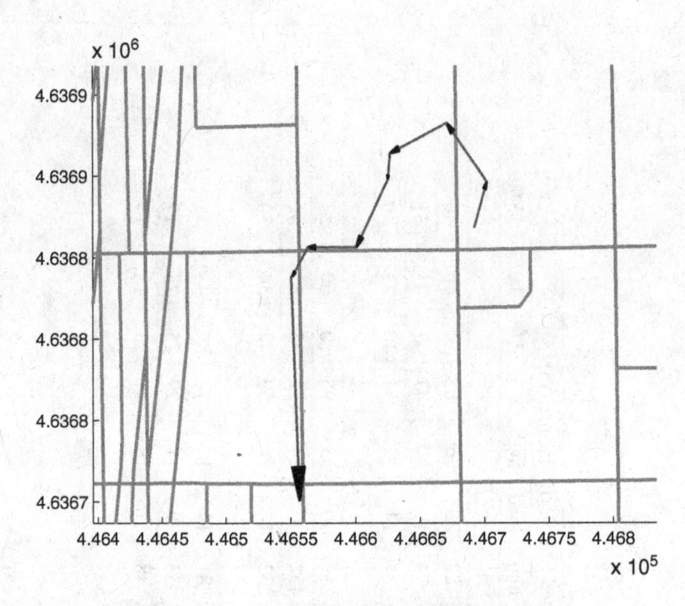

Fig. 7.1 A path with Fréchet distance greater than Hausdorff distance

and mapped onto the graph, where smaller distances are visualized in lighter shades and larger distances in darker shades. Such visual representation helps to identify areas in the constructed map that have larger distance to the ground-truth map.

7.3 Shortest Path-Based Distance

Another approach to the evaluation of constructed maps is using the shortest path-based distance proposed by Karagiorgou and Pfoser [7]. Here, a random set of corresponding origin-destination pairs is chosen in both the constructed and the ground-truth maps, and the shortest paths from origin to destination are computed in the respective maps. For an ideal map construction result, the shortest paths of the constructed map should match the paths in the ground-truth map. The larger the mismatch however, the larger is the dissimilarity between the maps. This experiment computes 500 shortest paths for randomly chosen pairs of origin and destination nodes. The discrete Fréchet distance and the average vertical distance are used to compare the resulting paths in the respective maps. The experiments use the ground-truth maps from the datasets that are presented in Chap. 5. A first impression on how different constructed maps affect respective shortest paths is given in Fig. 7.4. Given a specific origin and destination in the *Chicago* map, the computed shortest path is 3.66 km long in the ground-truth map (black dotted line). The respective shortest paths in the three generated maps are shown in red (solid, lighter line). In the map generated by Ahmed and Wenk's algorithm, the shortest path is 4.67 km

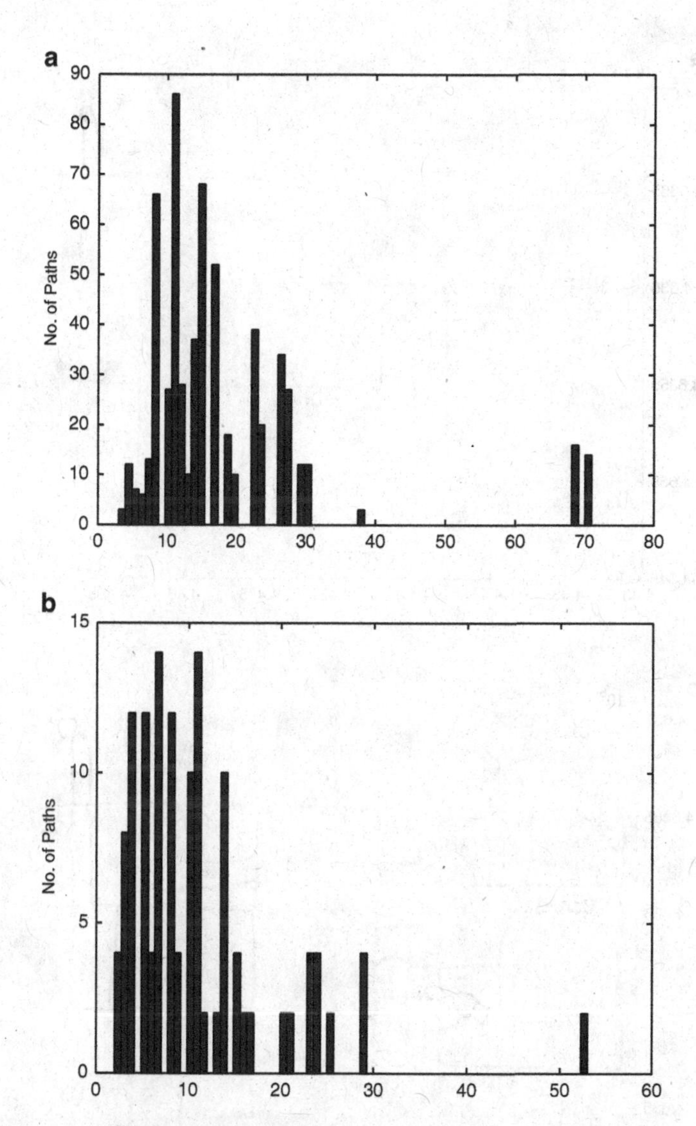

Fig. 7.2 Distributions of individual path distances for Biagioni—*Chicago*. (**a**) Path-based distance. (**b**) Directed Hausdorff distance

long (a discrete Fréchet distance with respect to the ground-truth map of 65 m and an average vertical distance of 21 m). The respective results for the other algorithms are 3.71 km (36 m, 5 m) for Biagioni and Eriksson, and 3.73 km (21 m, 5 m) for Karagiorgou and Pfoser. The shortest paths in the constructed maps are not always similar to the shortest paths in the respective ground-truth maps as in the case of Ahmed and Wenk (Fig. 7.4a), the resulting shortest path shows significant deviations. This result is in line with the path-based measure of Sect. 7.2, where also

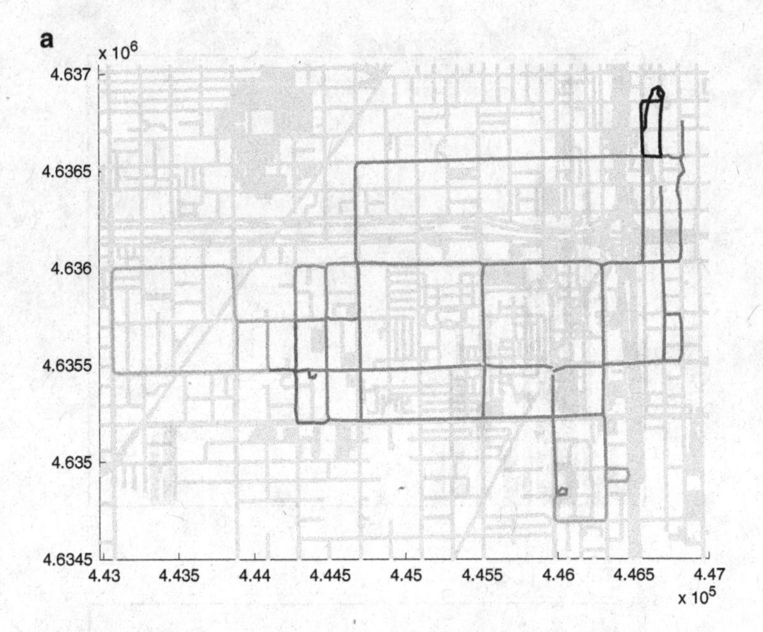

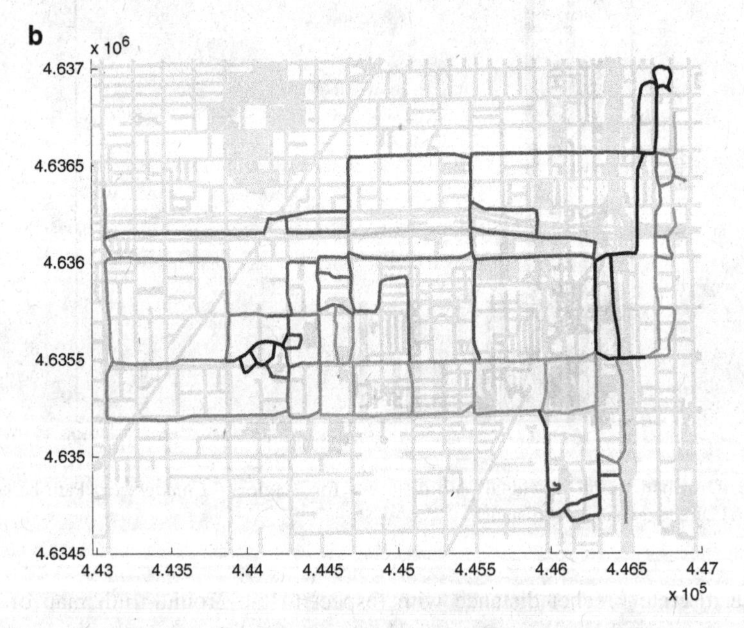

Fig. 7.3 Reconstructed graph overlayed on ground-truth map (*light gray*) for the *Chicago* dataset. Based on link-length three paths, edges in *lighter shades* have smaller path-based distance and *darker shades* have larger distance. (**a**) Biagioni. Plotted distances range between 3 m and 71 m. (**b**) Ahmed. Plotted distances range between 7 m and 201 m

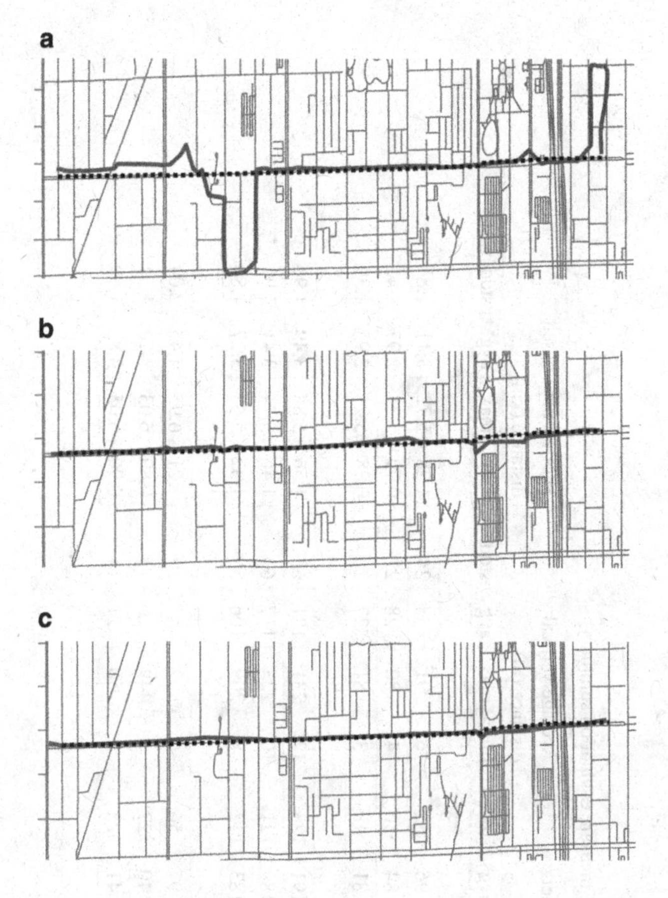

Fig. 7.4 Examples of shortest paths for the *Chicago* dataset. (**a**) Ahmed. (**b**) Biagioni. (**c**) Karagiorgou

Biagioni and Eriksson and Karagiorgou and Pfoser produced the best constructed maps.

The shortest path experiments are summarized in Table 7.2. The first column shows the percentage (%) of shortest paths that in each case could be computed, i.e., an algorithm might produce an accurate, but incomplete map and hence not all origin-destination pairs and thus shortest paths can be computed in this map. The second and third column show the two different distance measures (discrete Fréchet distance and average vertical distance) used to compare the resulting paths. The fourth column gives the min, max, and average length of the computed shortest

Table 7.2 Shortest path-based distance measure evaluation summary

Generated map	Found (%)	Discrete Fréchet distance (m)				Average vertical distance (m)				Shortest path-based distance (km)			
		min	max	avg	stddev	min	max	avg	stddev	min	max	avg	stddev
Athens													
Ahmed	97.6	13	234	96	62	6	91	38	24	1.28	5.72	3.11	1.84
Biagioni	94.2	7	214	84	50	4	80	28	21	0.79	5.23	2.97	1.41
Karagiorgou	96.8	7	212	81	48	3	81	27	20	0.78	5.21	2.95	1.39
Berlin													
Ahmed	93.2	21	469	191	123	12	231	121	63	1.56	5.88	3.49	1.96
Biagioni	91.8	20	461	189	121	10	223	117	60	1.46	5.74	3.41	1.97
Karagiorgou	93.8	18	428	183	112	8	209	106	58	1.32	5.67	3.27	1.84
Chicago													
Ahmed	99.8	13	208	97	56	6	92	43	19	1.21	6.95	4.45	2.04
Biagioni	98.6	4	98	40	27	2	49	20	13	0.89	6.03	3.76	1.57
Karagiorgou	99.2	4	103	41	28	2	50	21	14	0.90	6.05	3.82	1.59

paths. For example, for the *Berlin* data set, Ahmed and Wenk's algorithm constructs a map for which the generated shortest paths have a min, max, and average discrete Fréchet distance of 21 m, 469 m, and 191 m, respectively.

An aspect not captured by these distances are missing paths due to *limited map coverage*. Karagiorgou and Pfoser's algorithm constructs maps that have both good coverage and high path similarity. Consider for example the case of *Berlin*, which shows good coverage (93.8 %) and overall a small distance measures. This indicates similar paths and thus a constructed map similar to the ground-truth map.

Overall, shortest path sampling provides an effective means for assessing the quality of constructed maps as it not only considers *similarity*, but also the *coverage* of constructed maps.

7.4 Graph Sampling-Based Distance

The source code for computing the graph-sampling based distance has been provided by the authors of [5], and it was modified to use Euclidean distance as the data uses projected coordinate systems. The algorithm that computes this distance has four parameters: (1) *sampling density*, how densely the map should be sampled (marbles for generated map and holes for ground-truth map), 5 m is used; (2) *matched distance*, the maximum distance between a matched marble-hole pair, this distance is varied from 10 m to 120 m; (3) *maximum distance from root*, the maximum distance from a randomly selected start location for the traversal, 300 m is used; and (4) *number of runs*, the number of start locations to consider, 1000 is used. In the implementation, a seed location is selected in the plane and then a corresponding set of start locations for traversal in each graph is selected within a *matched distance* from the seed. Both graphs are traversed within *maximum distance from the root* to the seed, and then the union of all traversed edges is sampled based on the *sampling density* specified to produce a set of sample points. A larger *matched distance* might yield a larger number of sample points. To make the comparison of all generated maps consistent, a sequence of random locations for each dataset was generated and the first 1000 locations were used from the same sequence for each algorithm for which both maps (ground-truth and constructed) had correspondences within a *matched distance*. When two maps are very similar, they should have very few unmatched marbles and holes, which implies the precision, recall and F-score values should be very close to 1. In this case, as a superset of the ground-truth map was used, there should be a large number of unmatched holes, which implies lower recall and F-score values than in [5], but still the relative comparison of F-score values should provide an idea of whether an algorithm performs better than another.

The values for *matched distance* were chosen up to 120 m, to be consistent with the error associated with the input data. As mentioned in [5], some areas in the *Chicago* dataset show errors well above 100 m.

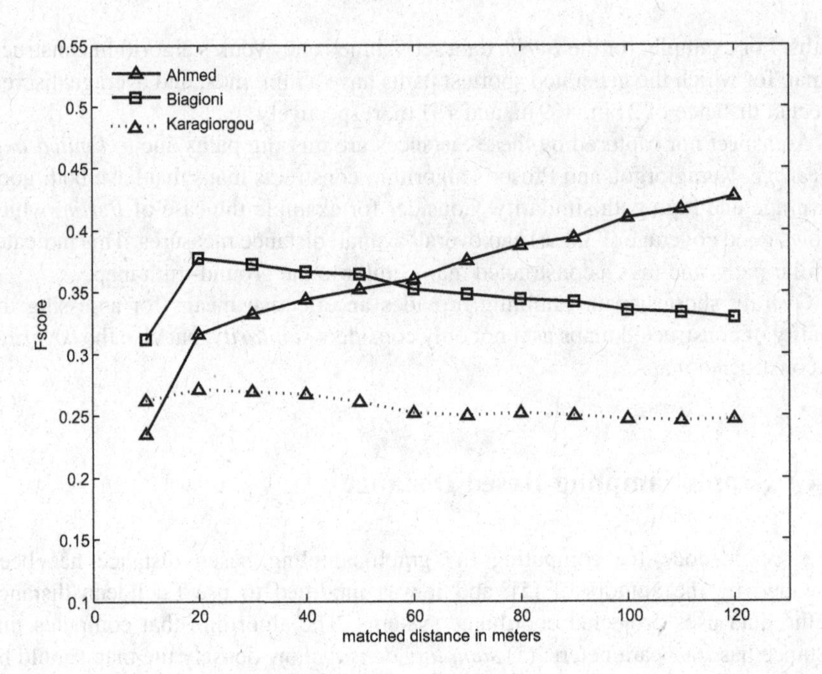

Fig. 7.5 Comparison of F-scores for the *Chicago* dataset

Figure 7.5 shows F-score values for the *Chicago* dataset for different generated maps. As the ground-truth is essentially a superset of the actual ground-truth covered by the tracking dataset, a larger matched distance decreases the number of unmatched marbles by matching these with available holes that probably are not part of the actual ground-truth. A higher recall value yields a higher F-score, which does not necessarily reflect better-quality maps (cf. maps in Chap. 5) such as in the case of Ahmed.

Figure 7.5 shows that the performance based on F-score declines for Biagioni and Eriksson and Karagiorgou and Pfoser with increasing *matched distance*. In investigating the reason of this unexpected behavior, it is found that although precision increases with increasing matched distance, the recall declines for these two algorithms; and smaller recall indicates a larger number of unmatched sample points in the ground-truth (empty holes). The constructed maps in Chap. 5 show that these two algorithms reconstruct fewer streets than other algorithms, which means they produce a smaller number of marbles to match with a larger number of holes. It was explained earlier in this subsection how the total number of holes might increase with the choice of a larger *matched distance*.

Hence, Table 7.3 shows the precision values instead of F-score and recall. According to precision values, the algorithms by Biagioni and Eriksson and by Karagiorgou and Pfoser perform best for dataset of *Chicago*, which is consistent with the findings using the other three distance measures.

Table 7.3 Precisions for *matched distance* 10 m, 40 m, 70 m, 100 m

Generated map	Precision value (for varying *matched distance*)			
	10	40	70	100
Athens				
Ahmed	0.265	0.442	0.503	0.579
Biagioni	0.450	0.586	0.662	0.727
Karagiorgou	0.343	0.489	0.561	0.647
Berlin				
Ahmed	0.123	0.326	0.422	0.485
Biagoni	0.239	0.510	0.551	0.586
Karagiorgou	0.294	0.590	0.633	0.649
Chicago				
Ahmed	0.312	0.563	0.658	0.738
Biagioni	0.491	0.699	0.730	0.775
Karagiorgou	0.602	0.740	0.751	0.801

Table 7.4 Local homology distance evaluation

Generated map	Bottleneck distance
Athens	
Ahmed	23.5
Biagioni	24.3
Karagiorgou	23.2

7.5 Local Homology Distance

The map construction algorithms were evaluated using the local homology distance for maps constructed from the *Athens* dataset. A disk cover of 19,845 disks of radius 100 m each was computed for the bounding box of the data. After computing the bottleneck distance between the persistence diagrams for the graphs in each disk, the average is computed as an overall measure. It is observed that based on the local homology distance, Karagiorgou is the closest to the ground-truth map with a distance of 23.2 m, Ahmed and Biagoni follow with distances of 23.5 and 24.2 m, respectively. Table 7.4 summarizes the results.

The local differences between two maps are illustrated using local distance signatures. These are computed by sampling each edge of one of the maps and creating a set of disks with sample points as centers, such that the union of disks covers the map. Then the local homology distance is computed for each disk, and each line segment on the graph is colored based on the average distance of the disks centered at the two endpoints of that line segment.

Figure 7.6 shows a comparison of a map constructed by the Karagiorgou and Pfoser algorithm (gray) with the map constructed by Biagioni and Eriksson algorithm (color-coded). The sampling density used to sample points from the

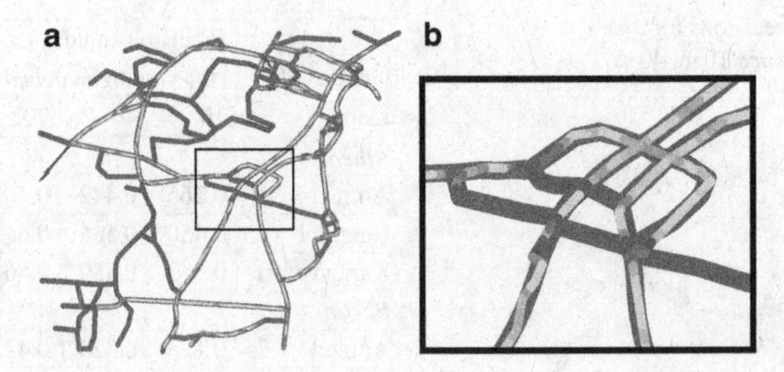

Fig. 7.6 Comparison of two generated maps of the *Athens* data set. The *thicker gray* map is Karagiorgou and the *color-coded* map is Biagioni. The *lighter shades* in the Biagioni map correspond to sections of the graph that have small signatures, and the *darker parts* correspond to large signatures. (**a**) Local homology distance. (**b**) Local homology distance, detailed view

Biagioni map is 5 m and the radius of each disk is 25 m. The lighter and darker shades represent smaller and larger distances, respectively. The local homology signature captures missing intersections, as illustrated in the detailed view of the signatures.

7.6 Summary

The best way to compare the constructed maps is in terms of coverage and accuracy. Here, it appears that density-based point clustering algorithms such as Biagioni and Eriksson's algorithm produce maps with lower complexity (fewer number of vertices and edges) and often fail to reconstruct streets that are not traversed frequently enough by the input trajectories. On the other hand, incremental track insertion algorithms such as Ahmed and Wenk's algorithm subsample all tracks to create a much denser output set, hence the complexity of their constructed maps is always higher. On the other hand, they fail to cluster tracks together when the variability and error associated with the input tracks is large. As a result, the constructed street maps contain multiple edges for a single street, which implies a larger constructed, but not necessarily more accurate road network.

In terms of map quality and accuracy, the maps reconstructed using the algorithms by Karagiorgou and Pfoser and by Biagioni and Eriksson generally have the smallest path-based and directed Hausdorff distances and their constructed maps can be considered more accurate. Although the algorithm by Ahmed and Wenk produces maps with good coverage and provide quality guarantees, their path-based distances are larger, since they employ less aggressive averaging techniques that would help cope with noise in the input tracks. In an effort to assess both accuracy and coverage, the shortest path-based distance shows for some algorithms good map quality, but at the same time only limited coverage. In this evaluation, Karagiorgou and Pfoser's produces maps that have both good coverage and high path similarity.

An overall observation to be made based on the experiments is that map construction algorithms tend to produce either accurate maps, or maps with good coverage, but not both. The algorithm of Karagiorgou and Pfoser, however seems to be a good compromise, in that it produces maps of good coverage and accuracy at the same time.

References

1. Ahmed, M., Wenk, C.: Constructing street networks from GPS trajectories. In: Proceedings of 20th Annual European Symposium on Algorithms, pp. 60–71 (2012)
2. Ahmed, M., Fasy, B.T., Wenk, C.: Local persistent homology based distance between maps. In: Proceedings of 22nd ACM SIGSPATIAL International Conference on Advances in Geographic Information Systems, pp. 43–52. ACM (2014)
3. Ahmed, M., Fasy, B.T., Hickmann, K.S., Wenk, C.: Path-based distance for street map comparison. ACM Trans. Spatial Algorithms Syst. **1**, 28 pp. (2015)
4. Alt, H., Guibas, L.J.: Discrete geometric shapes: matching, interpolation, and approximation - a survey. In: Sack, J.R., Urrutia, J. (eds.) Handbook of Computational Geometry, pp. 121–154. Elsevier, Amsterdam (1999)
5. Biagioni, J., Eriksson, J.: Inferring road maps from global positioning system traces: survey and comparative evaluation. Transp. Res. Rec. J. Transp. Res. Board **2291**, 61–71 (2012)
6. Biagioni, J., Eriksson, J.: Map inference in the face of noise and disparity. In: Proceedings of 20th ACM SIGSPATIAL International Conference on Advances in Geographic Information Systems, pp. 79–88 (2012)
7. Karagiorgou, S., Pfoser, D.: On vehicle tracking data-based road network generation. In: Proceedings of 20th ACM SIGSPATIAL International Conference on Advances in Geographic Information Systems, pp. 89–98 (2012)
8. OpenStreetMap. (2015) http://www.openstreetmap.org/

Chapter 8
New Directions

Abstract Map construction algorithms are useful beyond GPS-derived trajectory datasets. This chapter gives some examples of early-stage research towards novel applications. In recent years an ever increasing amount of social media data has become available. When considering the geospatial dimension of this data, using geocoded tweets for example, one can unveil location-based "stories" by means of geo-semantic extraction of a *network of interest* as described in this chapter. For eye tracking data, researchers use map construction techniques to recreate a user's "visual" trace based on positional snapshots. The result is used to study different optical representation concepts, such as distractions, content layout, and usability.

8.1 Social Media Tracking Data

An important resource in today's mapping efforts, especially for use in mobile navigation devices, is an accurate collection of point-of-interest (POI) data. However, by only considering isolated locations in current datasets, the essential aspect of how these POIs are connected is overlooked. The objective is to take the concept of POIs to the next level by computing *networks of interest* (NOIs) that encode different types of connectivity between POIs and that capture a person's type of movement and behavior while visiting these POIs [11]. This new concept of NOIs has a considerable application potential, including traffic planning, geomarketing, urban planning, and the creation of sophisticated location-based services, such as personalized travel guides and recommendation systems. Currently, the only datasets that consider connectivity of locations are road networks, which connect intersection nodes by means of road links purely on a geometric basis. POIs however, encode both geometric and semantic information (where and what) and it is not entirely obvious how to create meaningful links between them. Here, a *network of interest* is proposed to capture both *geometric* and *semantic* information by analyzing social media in the form of spatial check-in data. The concept of check-ins is used as a generic term for users actively volunteering their presence at a specific location. Existing road maps and POIs encode mostly geometric information and consist of street maps, but may also include subway maps, bus maps, and hiking trail maps. Such datasets can be derived from (GPS-based) *geometric trajectories* using map construction algorithms. This case also exploits so-called *behavioral trajectories*. They are obtained from social media in the form

M. Ahmed et al., *Map Construction Algorithms*, DOI 10.1007/978-3-319-25166-0_8

of spatial check-in data, such as geocoded tweets from Twitter. Similar to GPS tracking, the user contributes a *position sample* by checking in at a specific location. The main challenge arises from the fact that trajectories derived from geocoded tweets are typically quite sparse since individuals tend to publish their positions only at discrete occasions. However, by combining and analyzing time and location of such data, it is possible to construct event-based trajectories, which can then be used to analyze user mobility and to extract visiting patterns of places. The expectation towards behavioral trajectories is that by integrating them into a NOI, the resulting dataset will go beyond a homogeneous transportation network and will provide a means to construct an actual depiction of human interest and motion dependent on user context and independent of transportation means. Early maps were traces of people's movements in the world, depicting representations of people's experiences. NOIs on the other hand aim to fuse different qualities of such trace datasets obtained through intentional efforts (e.g., social media, Web logs) or unintentional efforts (e.g., routes from their daily commutes, check-in data) to provide for a *consequent modern map equivalent*.

In the following, a new algorithm is introduced to address the challenge of extracting a geosemantic network of interest from noisy, low-sampled geocoded tweets. The algorithm first segments the input dataset based on sampling rate and movement characteristics and then infers the respective network layers. To fuse the different network layers into a NOI, the concept of network hubs is used to align the different layers.

8.1.1 A Network of Interest

The goal is to construct a network of interest that reveals the *movement behavior* of users. This network of interest is a directed graph $G = (V, E)$, where the vertices V indicate important locations and the edges E important links between vertices according to observed user movements. In particular, two major aspects of the network of interest are interesting. The *geometric NOI aspect* provides a representation of how users actually move across various locations, thus preserving the actual geometry of the movement. The *semantic NOI aspect* represents the qualitative aspect of the network by identifying significant locations and links between them. These two aspects are treated as different layers of the same network of interest. In the following, the steps for constructing these layers and fusing them to produce the final network of interest are described.

8.1.2 Segmentation of Trajectories

Behavioral trajectories, as in this case derived from geocoded tweets, contain data to construct both the geometric and the semantic layer of a network of interest.

Conceptually, users tweet when they stroll around as well as when they commute in the morning. While all these tweets will result in behavioral trajectories, some of them depict actual movement paths, while others simply are tweets sent sparsely throughout the day. In what follows, the input data is separated into two subsets in order to extract the trajectories corresponding to the respective layer. The sparse subsets of the data are helpful in identifying significant locations and the denser subsets can be used to capture more fine-grained patterns of user movement. Users with frequent check-ins produce trajectories with high sampling rates, which provide a means to derive a geometric NOI layer. Low sampling rates on the other hand only allow to reason about abstract movement, which will be used to derive a semantic NOI layer.

With a focus on an urban scenario and transportation means, initially the mean speed is used to filter the trajectories and then the duration between samples is used to determine "abstract" and "concrete" movement. Figure 8.1a shows the trajectories classified by different sampling rates using the example of geocoded tweets for *London*. Using a heatmap coloring schema, concrete and abstract movements are shown in blue and red, respectively.

8.1.3 Geometric Layer Construction

The geometric NOI layer is constructed from frequently sampled trajectories. The algorithm follows a modified map construction approach (cf. [8, 9]) by (1) initially clustering position samples to derive network nodes, (2) linking nodes by means of edges derived from trajectories, and finally (3) refining the edge geometry.

To derive network nodes, the DBSCAN clustering algorithm [7] is used with a distance and a minimum-number-of-samples threshold. The segmented trajectories are used to create edges between the network nodes. The edges represent averaged trajectory geometries as two nodes can be connected by different trajectories. Finally, a reduction step is applied to simplify the constructed network. Edges of longer duration (difference of node timestamps) are reconstructed by using edges of shorter duration if their geometries exhibit similarity. Figure 8.2 shows an example of edges before and after the reduction step. Part of the larger geometry has been substituted with a more detailed geometry.

8.1.4 Semantic Layer Construction

The semantic NOI layer is constructed from trajectories with low sampling rates. Since such trajectories potentially cover large distances in between position samples, reconstructing the actual movement is difficult. By applying the DBSCAN clustering algorithm, a set of nodes is extracted that corresponds to the *hubs* of the semantic layer.

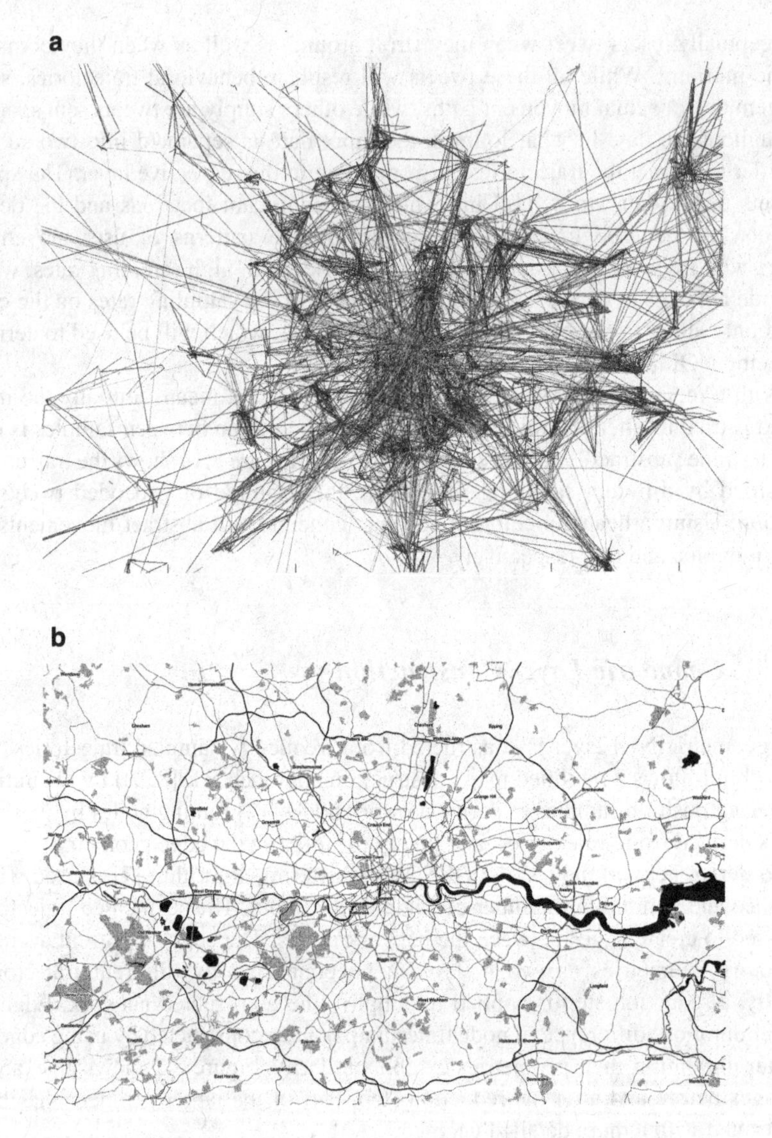

Fig. 8.1 Twitter trajectories and OSM network—*London*. (**a**) Twitter trajectories. *Blue* depicts "slow", *red* depicts "fast". (**b**) Respective OSM network (Color figure online)

Performing a linear scan of the trajectories reveals the respective trajectory portions that connect sets of nodes constituting hubs. Here, no reduction step is applied as the edge geometries of the semantic layer are too abstract, i.e., this layer represents a network with lower spatial accuracy, but with greater semantic value.

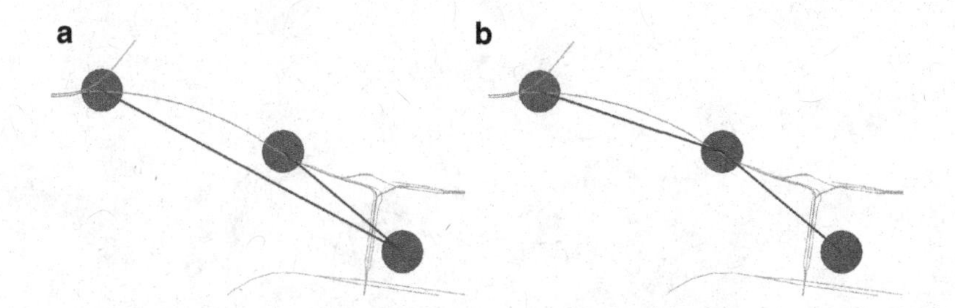

Fig. 8.2 Network reduction example. The constructed network is shown in *dark gray* and the underlying OSM road network in *light gray*. (**a**) Before reduction. (**b**) After reduction

8.1.5 Network Hubs

Hubs are POIs that users frequently depart from and arrive at. In particular, specific indicators for hubs are the number of constituting position samples stemming from many different users over extended time periods. DBSCAN is used to cluster relevant position samples using a distance and a minimum-number-of-samples threshold. The centroids of the resulting clusters are the candidate hubs. A candidate hub is included in the final set if it has a large number of incoming and outgoing edges at the same time. These conditions are used to ensure that the identified hubs correspond to places where a sufficiently large number of users frequently depart from and/or arrive at.

8.1.6 Layer Fusion

The final part of the process comprises the fusion of the geometric and semantic NOI layers. The NOI is constructed by starting with the semantic layer and merging the geometric layer into it. The intuition for this is that the semantic layer corresponds to a geometrically abstract, but semantically richer user movement that contains relevant transportation hubs. The geometric layer corresponds to a less semantic, but a more accurate depiction of movement, which can be used to fill in the gaps in the semantic layer. The fusion of these layers should result in a comprehensive movement network.

The fusion task involves (1) finding hub correspondences among the different network layers and (2) adding new links to the semantic layer for the uncommon portions of the NOI. Any remaining nodes of the geometric layer that have not been merged yet since they are not connected to the semantic layer are added in the end. A result of applying this conflation algorithm is shown in Fig. 8.3. Indicated are the circled hub correspondences between the semantic layer, the geometric layer, and the resulting fused network of interest.

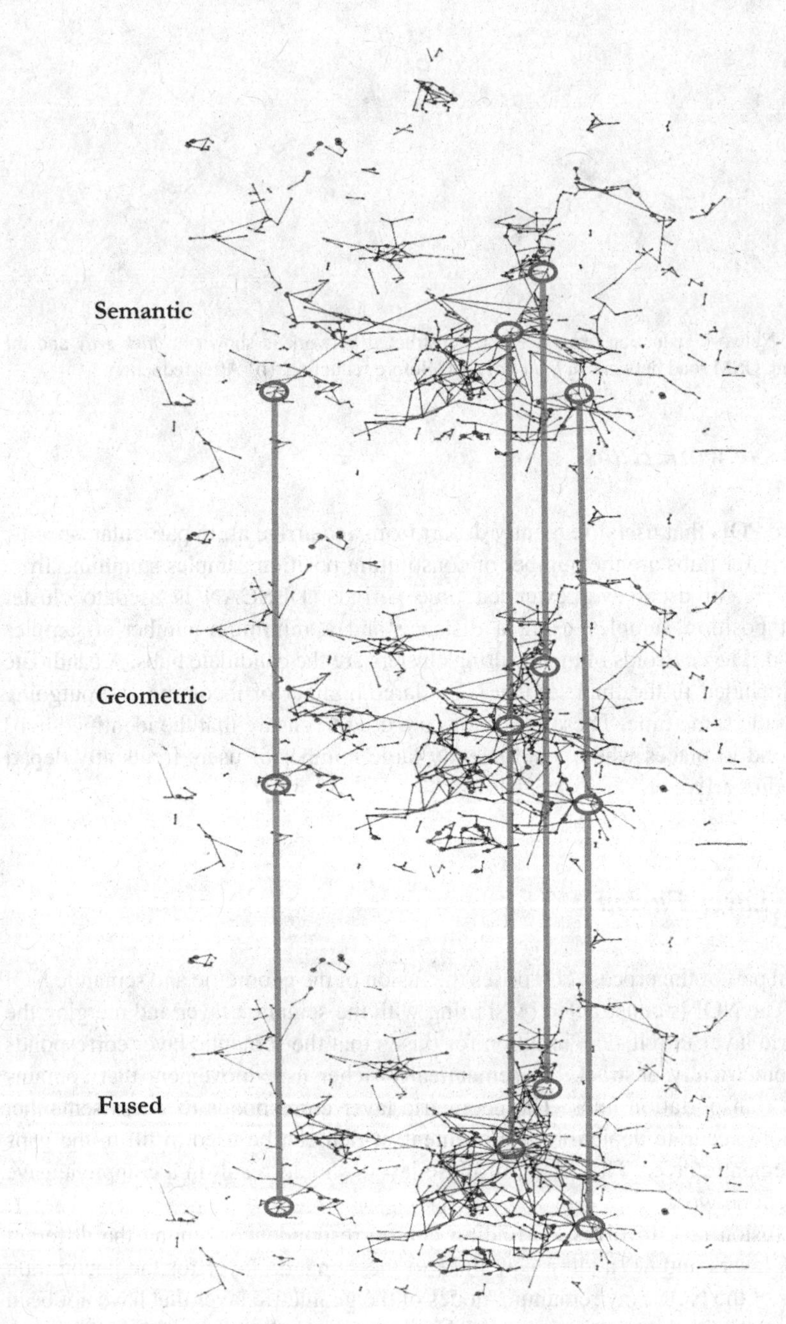

Fig. 8.3 Fused network—*London*

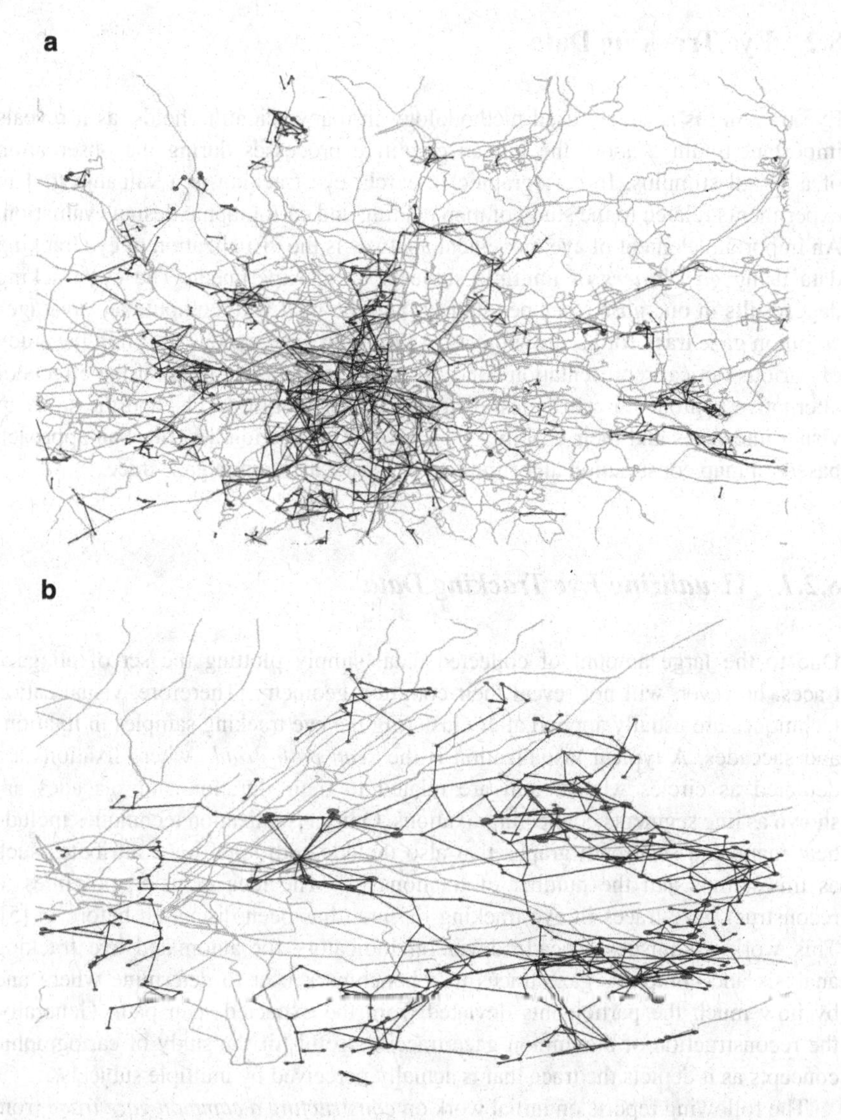

Fig. 8.4 Networks of interest. The constructed network is shown in *black* and the ground-truth network in *light gray*. (**a**) Network of interest—*London*. (**b**) Network of interest—*New York*

An overview of the quality of the constructed network of interest can be obtained by visual inspection, i.e., by comparing the network of interest to the ground-truth public transportation network and looking for similarities and differences. Figure 8.4 shows the NOIs of the cities of *London* (Fig. 8.4a) and *New York* (Fig. 8.4b). It is evident that, especially for the case of *New York*, the constructed NOI lines up well with the transportation network and identifies major hubs.

8.2 Eye Tracking Data

Eye tracking is a widely used methodology in many scientific fields, as it reveals important findings about the human cognitive processes during the observation of a visual stimulus. In cartographic research, eye tracking is a valuable tool in experiments related to the study of map reading and cartographic design evaluation. An important element of eye movement analysis is the visualization of eye tracking data using *gaze traces* of multiple subjects in an experiment. The eye tracking data results in one *gaze trace* per subject, and the goal is to compute an "average" common gaze trace for all the subjects. A common gaze trace is useful in the study of various optical representation concepts, such as the assessment of the effects of alternative contour line attributes, distractions, abstraction levels, and the study of visual interfaces and their usability in general. This section describes an approach based on map construction algorithms to construct a common gaze trace.

8.2.1 Visualizing Eye Tracking Data

Due to the large amount of collected data, simply plotting the set of all gaze traces, however, will not reveal their common geometry. Therefore, visualization techniques are usually applied after clustering the eye tracking samples in fixations and saccades. A typical visualization is the *scan path graph*, where fixations are depicted as circles whose radii are related to their duration and saccades are shown as line segments connecting fixations. Other visualization techniques include heat maps and scan path graphs that also include additional trace attributes such as timestamps and the number of fixations [4]. The idea of using polylines to reconstruct gaze traces in eye tracking research has been discussed before in [5]. This work establishes saccade deviation indicators for automated eye tracking analysis and compares gaze traces to a benchmark user to determine where and by how much the participants deviated from the expected scan path. Generally, the reconstruction of a common gaze trace is useful for the study of cartographic concepts as it depicts the trace that is actually perceived by multiple subjects.

The following reports on initial work on *constructing a common gaze trace* from multiple sequential raw eye tracking data samples [2]. The nodes of the constructed polyline contain information about the duration of fixations or other statistical values, which can also be attributed to line segments that represent saccadic movements. The motivation for the approach discussed in the following stems from map construction algorithms originally used to reconstruct road maps from GPS trajectories. Several such methods rely on trajectory clustering. Some of the algorithms in the literature [7, 12] operate on point data and do not take the temporal aspect into consideration. Others infer curved paths using k-means clustering of raw tracking data along with distance measures [6], or transform tracking data to discretized images using a data density function. These methods work well for frequently sampled and redundant tracking data [3], but are sensitive to noise. Other approaches, relying on computational geometry techniques [1] use highly accurate tracking data. The final category involves trace-clustering approaches that derive a connected road network from vehicle trajectories [8] of different movement types. The present approach [10] applies such a technique to eye tracking data to automatically extract "hubs" and to construct a polyline that corresponds to the observed geometry of cartographic lines.

8.2.2 Common Gaze Trace Construction

This common gaze trace construction algorithm takes eye tracking data obtained from user experiments as input and computes a common gaze trace represented by a polyline. Figure 8.5 shows the raw eye tracking data from three subjects, the contour line that the subjects have been asked to follow, and the raw individual gaze traces, one for each subject.

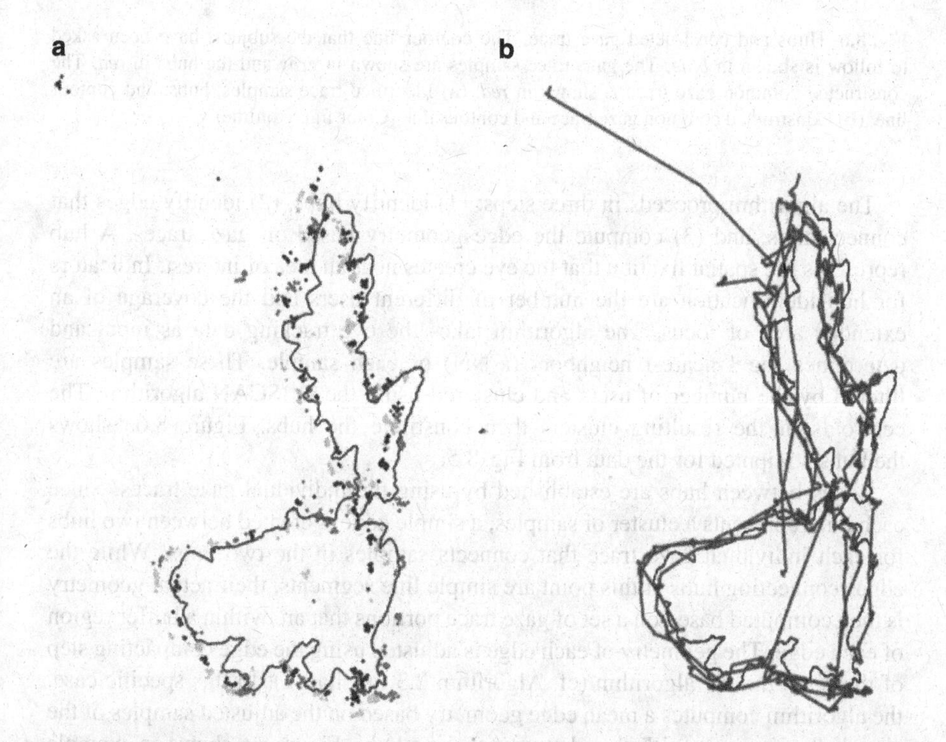

Fig. 8.5 Eye tracking data from three subjects. The contour line that the subjects have been asked to follow is shown in *blue*. The gaze samples are shown in three shades of *gray*, one for each subject. The individual gaze traces are shown in *red*. (**a**) Eye tracking samples and contour line. (**b**) Individual gaze traces and contour line (Color figure online)

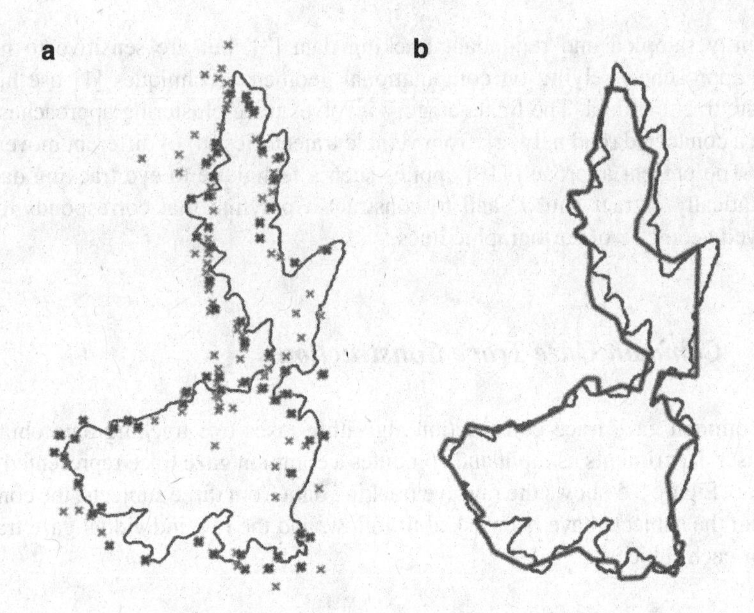

Fig. 8.6 Hubs and constructed gaze trace. The contour line that the subjects have been asked to follow is shown in *blue*. The gaze trace samples are shown in *gray* and the hubs in *red*. The constructed common gaze trace is shown in *red*. (**a**) Identified trace samples, hubs, and contour line. (**b**) Constructed common gaze trace and contour line (Color figure online)

The algorithm proceeds in three steps: (1) identify hubs, (2) identify edges that connect hubs, and (3) compute the edge geometry based on gaze traces. A hub represents the spatial fixation that the eye creates near an area of interest. Indicators for hub identification are the number of different users and the coverage of an extended area of focus. The algorithm takes the eye tracking data as input and determines the *k*-nearest neighbors (*k*-NN) of each sample. These samples are filtered by the number of users and clustered using the DBSCAN algorithm. The centroids of the resulting clusters then constitute the hubs. Figure 8.6a shows the hubs computed for the data from Fig. 8.5.

Edges between hubs are established by using the individual gaze traces. Since each hub represents a cluster of samples, a simple edge is created between two hubs for each individual gaze trace that connects samples in the two hubs. While the edges connecting hubs at this point are simple line segments, their actual geometry is then computed based on a set of gaze trace portions that are within a buffer region of each edge. The geometry of each edge is adjusted using the edge compacting step of the *TraceBundle* algorithm (cf. Algorithm 2.3 in Chap. 2). In this specific case, the algorithm computes a mean edge geometry based on the adjusted samples of the "bundle" of gaze traces that run between the two hubs. Figure 8.6 shows an example of the contour line that the subjects have been asked to follow, the extracted hubs (red crosses), and the constructed common gaze trace. What can be observed is that the constructed common gaze trace does not match the cartographic data in areas where no eye tracking samples are available.

References

1. Ahmed, M., Wenk, C.: Constructing street networks from GPS trajectories. In: Proceedings of 20th Annual European Symposium on Algorithms, pp. 60–71 (2012)
2. Bargiota, T., Mitropoulos, V., Krassanakis, V., Nakos, B.: Measuring locations of critical points along cartographic lines. In: Proceedings of 26th International Cartographic Conference (2013)
3. Biagioni, J., Eriksson, J.: Map inference in the face of noise and disparity. In: Proceedings of 20th ACM SIGSPATIAL International Conference on Advances in Geographic Information Systems, pp. 79–88 (2012)
4. Bojko, A.: Informative or misleading? Heatmaps deconstructed. In: Jacko, J. (ed.) Human-Computer Interaction. New Trends, Lecture Notes in Computer Science, vol. 5610, pp. 30–39. Springer, Berlin/Heidelberg (2009)
5. de Bruin, J.A., Malan, K.M., Eloff, J.H.P.: Saccade deviation indicators for automated eye tracking analysis. In: Proceedings of the 2013 Conference on Eye Tracking South Africa, pp. 47–54 (2013)
6. Edelkamp, S., Schrödl, S.: Route planning and map inference with global positioning traces. In: Computer Science in Perspective, pp. 128–151. Springer, Berlin (2003)
7. Ester, M., Kriegel, H.P., S, J., Xu, X.: A density-based algorithm for discovering clusters in large spatial databases with noise. In: Proceedings of 2nd International Conference on Knowledge Discovery and Data Mining, pp. 226–231 (1996)
8. Karagiorgou, S., Pfoser, D.: On vehicle tracking data-based road network generation. In: Proceedings of 20th ACM SIGSPATIAL International Conference on Advances in Geographic Information Systems, pp. 89–98 (2012)
9. Karagiorgou, S., Pfoser, D., Skoutas, D.: Segmentation-based road network construction. In: Proceedings of 21st ACM SIGSPATIAL International Conference on Advances in Geographic Information Systems, pp. 450–453 (2013)
10. Karagiorgou, S., Krassanakis, V., Vescoukis, V., Nakos, B.: Experimenting with polylines on the visualization of eye tracking data from observations of cartographic lines. In: Proceedings of 2nd International Workshop on Eye Tracking for Spatial Research (2014)
11. Karagiorgou, S., Pfoser, D., Skoutas, D.: Geosemantic network-of-interest construction using social media data. In: Duckham, M., Pebesma, E., Stewart, K., Frank, A.U. (eds.) Geographic Information Science. Lecture Notes in Computer Science, vol. 8728, pp. 109–125. Springer International Publishing, New York (2014)
12. Zhang, T., Ramakrishnan, R., Livny, M.: Birch: An efficient data clustering method for very large databases. In: Proceedings 1996 ACM SIGMOD International Conference on Management of Data, pp. 103–114 (1996)

References

Chapter 9
Resources

Abstract This chapter introduces resources that complement the scientific discussion of map construction algorithms and provide the interested researcher with the simplest possible means to start experimenting with map construction algorithms. The Map Construction Web Portal and its content are briefly discussed and user guides are provided for several map construction algorithms.

9.1 Map Construction Web Portal

The ambition of this book is to provide the reader with an introduction to map construction algorithms. Since map construction is a very active research field, this book can only capture a snapshot of the start-of-the-art in this field. To stand the test of time, the authors have established the web site http://www.mapconstruction.org/ as a repository for map construction data and algorithms, and other researchers are invited to contribute by uploading code and benchmark data supporting their map construction algorithms. The expectation is that such a central repository will encourage a culture of sharing and will enable the development of improved map construction algorithms.

Currently the site (cf. Fig. 9.1) contains a list of map construction papers, source code, and data. The list of map construction algorithms includes links to the papers, presentations and also source code. The authors also make the three new benchmark datasets (*Chicago*, *Berlin*, *Athens*), map construction outputs (visualizations) as well as the source code for various evaluation measures available.

9.2 User Guides

What follows are some brief notes on how to use the algorithms by Ahmed and Wenk [1], Biagioni and Eriksson [2], and Karagiorgou and Pfoser [3]. This should allow the interested reader to repeat the experiments discussed in this book.

M. Ahmed et al., *Map Construction Algorithms*, DOI 10.1007/978-3-319-25166-0_9

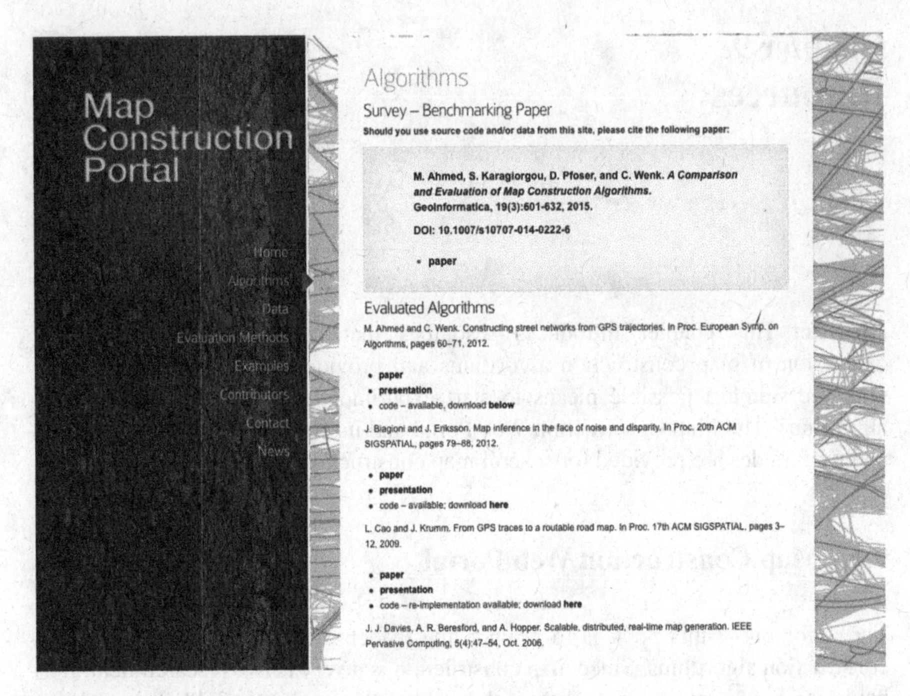

Fig. 9.1 Map construction portal web page

9.2.1 *Ahmed and Wenk*

The algorithm by Ahmed and Wenk [1] has been implemented in Java, using a projected coordinate system for the trajectories. The implementation includes partial map-matching and insertion of new edges, but not the minimum-link averaging of existing edges. The code constructs an undirected embedded graph as output, and if altitude information is available for the position samples it is able to produce non-planar graphs.

Input file format

The algorithm accepts the following input formats.

Format 1: The `HAS_ALTITUDE` parameter should be set to `FALSE` for this input format type. The input trajectories are given as `<x y timestamp>`.
Example:

```
482785.9 4216659.1 49039.0
483396.7 4216956.2 49069.0
```

Format 2: The `HAS_ALTITUDE` parameter should be set to `TRUE` for this input format type. The input trajectories are given as `<x y z timestamp>`.

Example:

```
482785.9 4216659.1 0.0 49039.0
483396.7 4216956.2 0.0 49069.0
```

Output file format

The constructed map output consists of two files: `vertex.txt` and `edges.txt`.

The vertex file format is `<vertexid,x,y,z>`.
Example:

```
0, 482785.9, 4216659.1, 0.0
1, 483396.7, 4216956.2, 0.0
2, 483693.2, 4216953.1, 0.0
```

The edge file format is `<edgeid,vertexid1,vertexid2>`.
Example:

```
1, 0, 0
2, 0, 234
3, 1, 127
```

Running the program

1. Download the code from http://www.mapconstruction.org/.
2. In order to run the code one has to choose the following parameters:
   - `INPUT_PATH`—path to the folder which has the trajectory files.
   - `OUTPUT_PATH`—path where the output will be written to.
   - `EPS`—epsilon in meters.
   - `HAS_ALTITUDE`—true if the trajectory file has altitude information, false otherwise.
   - `ALT_EPS`—minimum altitude difference in meters to be identified as two different streets.
3. After choosing the parameters, the program itself can be executed in two ways using Java.

   Option 1: Execute `script_to_run.sh` directly from a shell. Edit this file to choose parameters.

   Option 2: Import `MapConstruction` as a project in Eclipse and run it by passing the parameters as program arguments in the following order: `INPUT_PATH OUTPUT_PATH EPS HAS_ALTITUDE ALT_EPS`.

Contacts

For questions and bug reports, please email Mahmuda Ahmed (mahmudaahmed@gmail.com) or Carola Wenk (cwenk@tulane.edu).

9.2.2 Biagioni and Eriksson

The algorithm by Biagioni and Eriksson [2] has been implemented as a pipeline in Python. The implementation includes only density estimation, skeleton computation, and topology refinement, but not the final geometry refinement. Each step of the pipeline produces a different intermediate map construction result, with the final output being a directed embedded graph.

Input file format

The code takes latitude/longitude coordinates as input, and the trajectory files need to be in the format `<id,lat,lng,timestamp,previous_id, next_id>`.
Example:

```
1, 52.514445, 13.389166, 2262324.0, None, 2
2, 52.514168, 13.383055, 2262386.0, 1, 3
3, 52.513615, 13.380001, 2262459.0, 2, 4
```

Output file format

The output of this algorithm consists of one file, and an edge is represented as a sequence of latitude longitude pairs. An empty line represents the end of an edge. The output file is in the following format: `<lat,lng>`.
Example:

```
52.5112871499, 13.3867129606
52.5111781395, 13.3867512188

52.5111781395, 13.3867512188
52.5112871499, 13.3867129606
```

Running the program

The following Python libraries have to be installed in order to run the code for this algorithm: cython, numpy, scipy, PIL (or Pillow), opencv and rtree.

1. Download the code from http://www.cs.uic.edu/bin/view/Bits/Software.
2. Create density (kde.png) from trips:
```
python kde.py -p
 trips/02_GPX_Tracks_valid_flat_text_james/
```
3. Create gray-scale skeleton (skeleton.png) from the density:
```
python skeleton.py kde.png skeleton.png
```
4. Extract map database (skeleton_maps/skeleton_map_1m.db) from grayscale skeleton
```
python graph_extract.py skeleton.png
  bounding_boxes/bounding_box_1m.txt
  skeleton_maps/skeleton_map_1m.db
```

5. Map-match trips onto map database
```
python graphdb_matcher_run.py -d
  skeleton_maps/skeleton_map_1m.db -t
  trips/02_GPX_Tracks_valid_flat_text_james/ -o
  trips/matched_02_GPX_Tracks_valid_flat_text_james/
```
6. Prune map database with map-matched trips, producing pruned map database (skeleton_maps/skeleton_map_1m_mm1.db)
```
python process_map_matches.py -d
  skeleton_maps/skeleton_map_1m.db -t
  trips/matched_02_GPX_Tracks_valid_flat_text_james/
  -o skeleton_maps/skeleton_map_1m_mm1.db
```
7. Refine topology of pruned map, producing refined map (skeleton_maps/skeleton_map_1m_mm1_tr.db)
```
python refine_topology.py -d
  skeleton_maps/skeleton_map_1m_mm1.db -t
  skeleton_maps/skeleton_map_1m_mm1_traces.txt -o
  skeleton_maps/skeleton_map_1m_mm1_tr.db
```
8. Map-match trips onto refined map
```
python graphdb_matcher_run.py -d
  skeleton_maps/skeleton_map_1m_mm1_tr.db -t
  trips/02_GPX_Tracks_valid_flat_text_james/ -o
  trips/matched_02_GPX_Tracks_valid_flat_text_james_mm1_tr/
```
9. Prune refined map with map-matched trips, producing pruned refined map database (skeleton_maps/skeleton_map_1m_mm2.db)
```
python process_map_matches.py -d
  skeleton_maps/skeleton_map_1m_mm1_tr.db -t
  trips/matched_02_GPX_Tracks_valid_flat_text_james_mm1_tr/
  -o skeleton_maps/skeleton_map_1m_mm2.db
```
10. Convert pruned refined map database to text file (final_map.txt)
```
python streetmap.py graphdb
  skeleton_maps/skeleton_map_1m_mm2.db final_map.txt
```

Contacts

For questions and bug reports, please email James Biagioni (jbiagi1@uic.edu) or Jakob Eriksson (jakob@uic.edu).

9.2.3 *Karagiorgou and Pfoser*

The algorithm by Karagiorgou and Pfoser [3] has been implemented in Matlab.

Input file format

The code takes projected coordinates as input and the input files need to be in the following format: `<x y timestamp>`.
Example:

```
483389.0 4207889.6 47019.0
483422.3 4207877.3 47049.0
```

Output file format

The output of this algorithm consists of two files `tracebundle_vertices.txt` and `tracebundle_edges.txt`. The vertex file format is `<vertexid,x,y>`.
Example:

```
1, 484682.083645, 4216742.764901
2, 484795.314682, 4216860.778676
3, 484657.964168, 4216610.040074
```

The edge file format is `<edgeid,vertexid1,vertexid2>`.
Example:

```
1, 458, 409, 1
2, 1, 458, 0
3, 3, 8, 0
```

Running the program

1. Download the code from http://www.mapconstruction.org/.
2. The source code lies in the `/source` directory.
3. Add the `/source` and the `/libraries` directories to the current working path of MATLAB.
4. Run the `intersection_nodes_extraction.m` file.
5. Run the `tracebundle.m` file.

Contacts

For questions and bug reports, please email Sophia Karagiorgou (karagior@imis.athena-innovation.gr or Dieter Pfoser (dpfoser@gmu.edu).

References

1. Ahmed, M., Wenk, C.: Constructing street networks from GPS trajectories. In: Proceedings of 20th Annual European Symposium on Algorithms, pp. 60–71 (2012)
2. Biagioni, J., Eriksson, J.: Map inference in the face of noise and disparity. In: Proceedings of 20th ACM SIGSPATIAL International Conference on Advances in Geographic Information Systems, pp. 79–88 (2012)
3. Karagiorgou, S., Pfoser, D.: On vehicle tracking data-based road network generation. In: Proceedings of 20th ACM SIGSPATIAL International Conference on Advances in Geographic Information Systems, pp. 89–98 (2012)

Index

M. Ahmed et al., *Map Construction Algorithms*, DOI 10.1007/978-3-319-25166-0

Printed in the United States
By Bookmasters